Perspectives for
Geothermal
Energy in Europe

Perspectives for
Geothermal
Energy in Europe

Editor

Ruggero Bertani

Enel Green Power, Italy

World Scientific

W JERSEY · LONDON · SINGAPORE · BEIJING · SHANGHAI · HONG KONG · TAIPEI · CHENNAI · TOKYO

Published by

World Scientific Publishing Europe Ltd.

57 Shelton Street, Covent Garden, London WC2H 9HE

Head office: 5 Toh Tuck Link, Singapore 596224

USA office: 27 Warren Street, Suite 401-402, Hackensack, NJ 07601

Library of Congress Cataloging-in-Publication Data

Names: Bertani, Ruggero, editor.
Title: Perspectives for geothermal energy in Europe / edited by
    Ruggero Bertani (Enel Green Power, Italy).
Description: New Jersey : World Scientific, [2017]
Identifiers: LCCN 2016039136 | ISBN 9781786342317 (hc : alk. paper)
Subjects: LCSH: Geothermal energy--Europe.
Classification: LCC GB1199.8.E85 P47 2017 | DDC 333.8/8094--dc23
LC record available at https://lccn.loc.gov/2016039136

British Library Cataloguing-in-Publication Data
A catalogue record for this book is available from the British Library.

Desk Editors: Chandrima Maitra/Mary Simpson/Shi Ying Koe

Typeset by Stallion Press
Email: enquiries@stallionpress.com

Printed in Singapore

About the Authors

Ruggero Bertani received a degree in Physics from Pisa University in 1979. He is Project Director in the Innovation Department; working in the Enel Green Power since 1982. Bertani is a Coordinator and responsible of the DESCRAMBLE H2020 project. He has been the Executive Director for six years since 1998 of the International Geothermal Association (IGA). He is also the main author of the geothermal chapter of the IPPC report 2009 and leading author of the geothermal chapter in the Special Report on the Renewable Energy Sources (SRREN) of IPCC working group. He is also the President of European Geothermal Energy Council (EGEC) and President of European Technological Innovation platform (ETIP) for Deep Geothermal in the SET Plan.

Philippe Dumas holds a Master's degree in European Affairs (1999). He first worked in a European engineering company as representative in Brussels for EU affairs (2000–2007). He was involved in geothermal energy with EGEC, being project manager for European projects. Since September 2008, he is the EGEC Secretary General in Brussels managing the association. He is the author and co-author of several publications; frequent contributor to conferences, workshops and seminars; teacher at University of Marseille on European lobbying; active in a number of EU-funded research and promotion projects from 2000 until today.

Joseph Bonafin holds a Master's degree in Mechanical Engineering, with specialization in power generation systems. His background in the Organic Rankine Cycle technology started with a heat recovery project at Wartsila. In 2009, he joined Turboden. After eight years of extensive geothermal business experience, he currently manages the geothermal activities as sales and business development leader at Turboden. He is involved in all the project phases, from early technical development to finalization of the commercial negotiation. He currently follows geothermal projects globally with specific expertise in the power plant development.

Ólafur G. Flóvenz has been active in geothermal research since 1974. He holds a Dr. Scient. in Geophysics from the University of Bergen, Norway. He has been the General Director of Iceland Geo-Survey (ÍSOR) and its predecessor, the GeoScience division of the National Energy Authority (NEA) of Iceland since 1997. He has had a permanent position as a geothermal expert since 1979, and as the head of the geophysical department of NEA since 1985. He has been an Adjunct Professor in Geothermics at the University of Iceland. He has also been on numerous boards of directors, including that of the International Geothermal Association (IGA) from 2001 to 2007, Nordic Volcanological Institute and the Research Council of Iceland. His main field of expertise covers geophysical and geothermal exploration.

Brynja Jónsdóttir has been working in the geothermal sector since 2001. First for the GeoScience division of the National Energy Authority (NEA) of Iceland and since 2003 in publicity and public relations for Iceland GeoSurvey (ÍSOR). She has a B.Sc. in Geography and a B.A. in German, both from the University of Iceland, as well as a Diploma in Marketing and Public Relations from the University of Reykjavik. She is currently studying towards a Master's degree in Practical Dissemination for Culture.

Adele Manzella is Senior Scientist at the National Research Council of Italy (CNR) and works as a geophysicist in geothermal exploration to conduct field and theoretical investigations of geothermal systems in Italy and abroad. Her main research interests are the integration of different geothermal exploration methods for reservoir characterization, and feasibility studies for geothermal plants. She coordinated for CNR, the Italian geothermal assessment projects of Southern Italy, and led most of the CNR participation to EU projects dedicated to geothermal energy, regarding exploration methods development, coordination of research efforts and geothermal networking, and promotion and support for the development of geothermal energy.

Assunta Donato is a research fellow at the National Research Council of Italy (CNR) and a PhD student at the University of Florence. Her ongoing research is aimed at geothermal resources assessment by integration of geological, geophysical and geochemical data, environmental and regulative aspects of geothermal energy.

Gianluca Gola received his PhD in Earth Sciences, is a Geophysicist and works as researcher at the National Research Council of Italy (CNR). The research activity includes the study of thermo-physical parameters of the rocks and the numerical modeling of geological processes with a focus on the thermal and rheological aspects.

Alessandro Santilano is a research fellow at the National Research Council of Italy (CNR). His ongoing research is aimed at the electro-magnetic geophysical study and 3D integrated modeling of geothermal systems. He is currently a PhD student at the Polytechnic of Turin where he is testing probabilistic optimization algorithms for the analysis of electromagnetic geophysical data.

Eugenio Trumpy is a PhD in Earth Sciences, a geologist and works as technologist at the National Research Council of Italy (CNR). His main research interests include geological, geothermal and geophysical digital data management, Geographical Information Systems (GIS), Spatial Data Integration and analysis, 3D geological modeling, business intelligence tools applied on geothermal data. Since 2007 he is in charge of the maintenance, design and development of the Italian Geothermal Database and in 2014 he coordinated the implementation of the European Information Platform pilot project. Since 2015 he leads the dissemination activity in European geothermal projects for CNR.

Sakir Simsek is a Professor of the Geological (Hydrogeology) Engineering Department of the Hacettepe University, Ankara,

Turkey. Director of the special environmental protection project of Pamukkale thermal springs and travertines as World Cultural Heritage Project. He is the Principal Investigator of IAEA Chemical and Isotopic Survey Projects and JICA projects on geothermal exploration in Turkey, and currently a member of the Board of Turkish Geothermal Association (TGA) and a member of the Board of Directors of the International Geothermal Association (IGA) for 2001–2004 and 2004–2008.

Jan-Diederik van Wees holds a PhD in tectonics. He is the principal scientist in geothermal research at TNO, and Professor at Utrecht University. He has published over 60 papers in leading international journals on tectonics, reservoir engineering, resource assessment, and techno-economic models. Van Wees is sub-program manager exploration in the Joint Program on Geothermal Energy of the European Energy Research Alliance, and Vice President of the European Technological Innovation platform (ETIP) for Deep Geothermal in the SET Plan. He has a scientific coordinating role in the IMAGE FP7 project, and GEMEX H2020 project.

Maarten Pluymaekers obtained his Master's degree in Geology at the Utrecht University in 2007. Subsequently he joined TNO as a geologist, and is an expert in characterization and prediction of geothermal reservoir quality. Recent projects in the field of geothermal energy includes the development of ThermoGIS (a national information system for geothermal exploration based on technical-economical assessment), development of DoubletCalc (a geothermal reservoir simulator) and several local and regional geothermal potential assessments for both clastic and carbonate reservoirs.

Hans Veldkamp is currently employed by the Netherlands Organisation for applied scientific research (TNO) as senior geoscientist. He also holds the position of Treasurer of the board for the Dutch Geothermal Platform. He has an MSc in Structural and Sedimentary Geology from the Vrije Universiteit of Amsterdam (1991). He has extensive experience in static modeling of the shallow and deep subsurface using seismic and well data, and geostatistics. He is currently involved in various European projects such as IMAGE (geothermal exploration), SURE (radial jetting) and DeStress (soft stimulation). For the latter, he is acting as WP leader.

Serge van Gessel holds a Master's degree in Geology at Utrecht University (1994). He is currently working as a senior geoscientist and advisor for the Ministry of Economic Affairs and the Ministry of Infrastructure and Environment at the Geological Survey of the Netherlands (TNO). Besides his day-to-day work at TNO, Serge holds the position of Chair of the Geo-Energy Expert Group at Euro-GeoSurveys (European Geological Surveys).

Damien Bonté holds a PhD degree from Utrecht University, with specialization on thermal characterization in relation with large-scale geodynamic processes and magmatism. In 2014, he re-joined Utrecht University as a postdoctoral researcher, after a couple of years in geothermal consulting, to work on the GEOCAP project (Indonesia-NL geothermal), providing courses and research. He is now working on GEMex (Mexico-EU research collaboration) as a researcher and WP leader.

Ladislaus Rybach is of Hungarian origin, Emeritus Professor of Geophysics at ETH Zurich, co-founder, now Scientific Advisor of ETH spin-off company GEOWATT AG Zurich. He is active world-wide as expert and lecturer, President (2007–2010) of the International Geothermal Association (IGA), co-founder of the IEA Geothermal Implementing Agreement and Former Executive Committee Chairman; Honorary Doctor and Professor of Eötvös University, Budapest.

Burkhard Sanner is a Geologist, expert on shallow geothermal applications, working for UBeG GbR in Germany since 2004 and for Justus-Liebig-University Giessen most of the time from 1981 to 2004. In 1985, he was the project manager for the first German test plant for ground source heat pumps (Schwalbach GSHP Research Station of Helmut Hund GmbH), involved in IEA-projects on GSHP and UTES from 1986 to 2004 and in EU-projects on these topics thereafter. He was the co-founder of the European Geothermal Energy Council (EGEC) in 1998 and President of EGEC from 2004 to 2016;

board member of the European Technological and Innovation Platform (ETIP) for Renewable Heating and Cooling; Chairman (1998–2004) and since then Vice Chairman of the committee for German guideline VDI 4640 on shallow geothermal technology.

Luca Angelino joined the European Geothermal Energy Council in 2011, where he is currently Head of Policy and Regulation. He monitors and provides strategic advice on the political, regulatory, and economic environment relevant to the sector; and is Project Officer for the EU-funded projects GEOELEC, GEODH and REGEOCITIES. Since 2011 he has been coordinating the collection and analysis of data on geothermal market trends for the annual EGEC market report. Prior to joining EGEC, Luca worked for other trade associations in the energy sector, a think tank, and the European Commission.

Contents

Introduction

Ruggero Bertani

Enel Green Power, Via A. Pisano 120
56122 Pisa, Italy
ruggero.bertani@enel.com

1. Historical Notes

Throughout history there has been continuous contact between humans and geothermal phenomena; volcanoes, fumaroles, hot springs, steaming grounds and geysers, and all other manifestations of thermal content. The first indirect benefits from geothermal energy were related to domestic uses, in hot-water bathing, cooking and the use of obsidian by-products for the creation of simple tools. According to Cataldi *et al.* (1999), the first stage in the relationship between humanity and geothermal can be called "year zero of geothermics," dating from the Paleolithic era to the beginning of the Metal Age, when natural, simple human settlements lived. During this time, different stages of utilization of geothermal products were identified. Additionally, the inference between geothermal sources and their subterranean force/divinity was a clear part of people's belief system at the time. Whereas "strong" phenomena such as volcanic eruptions and related earthquakes caused fear within populations, it is easy to imagine that people made good use of "soft" geothermal sources such as hot springs.

1.1. Mediterranean Region

Let's look around the Mediterranean countries: in Italy, the geothermal phenomena are abundant, and the utilization was strongly linked with the settlements nearby the thermal manifestations, with the construction of a circular pond (*tholos*, R.G. Cremonesi in Cataldi *et al.* [1999]), and an extensive mining of obsidian and other volcanic raw materials.

The Etruscans (12th century B.C.) in Tuscany can be considered as the precursor of the geothermal industry, using the minerals naturally present in the Boraciferous region of Tuscany, inside the triangular area between Pisa, Siena, and Grosseto: alum, borates (the manifestation are known as "soffioni boraciferi"), kaolin, iron oxides, sulfur, and travertine. Not only did they exploit the hydrothermal ores, but they also developed a very sophisticated technique for making high-quality products like pottery and many others (Cataldi *et al.*, 1999).

But it was with the beginning of the Roman Empire (27th century B.C. to 5th century A.D.) that the use of spa (*thermae*) along the entire Mediterranean basin, as well as the use of geothermal by-products, following and improving the Etruscan legacy, producing and disseminating cement slurries made by hydrothermal minerals (pozzolan), travertine, other volcanic material for buildings, kaolin for ceramics and textile industries, boron compounds and iron oxides for pottery paintings, and masses of smectic clays rich in hydrothermal minerals for therapeutic use (Cataldi *et al.*, 1999) came into human life.

The Middle Ages was a long period of decadence in technology and life quality, when geothermal energy utilization in the former Western Roman Empire was strongly reduced, whereas in the Byzantine, Arab and Ottoman civilizations in the Eastern part of the world, thermal water utilization was continuously preserved.

Only after the 11th century, in Tuscany, did the mining activities restart in the Boraciferous region, mainly for alum production, which was used largely in the textile (wool processing) industry: the economical importance of the area become so high that a war between Florence and Volterra in 1472 A.D. ended with the final control of the Florentine craft guild for wool (Arte della Lana, P.D. Burgassi in Cataldi *et al.*, 1999).

1.2. Rest of Europe

In the Precarpathian and Pannonia basin area (Austria, Slovakia, Romania and Hungary), thermal springs are abundant and widespread (Cohut *et al.* and Fendek *et al.* in Cataldi *et al.*, 1999). Thermal baths have been used in a large number of sites, such as Baden (Austria), Piešt'any (Slovakia), Budapest (Hungary), and Oradea (Romania). During the late Middle Ages, the use of geothermal for medical purposes was developed, and it was further increased under the Ottoman Empire (thermal station locally known as Kaplica).

In Poland (Sokolowski *et al.* in Cataldi *et al.*, 1999), only a moderate terrestrial heat flux is present, and only a few thermal springs have been utilized; the site of Chiechocinek in the middle of the country had low-temperature curative water for about 10 centuries. The most important modern utilization is the zakopane complex, on the Carpathians.

In France (Gibert *et al.* in Cataldi *et al.*, 1999), Chaudes-Aigues (hot waters) represent the presence of important hot spring locations from the Middle Ages: it is one of the hottest European thermal waters at 82°C! It has been in use since 14th century, and it can be considered as the first example of district heating in the world, using a wooden piping system.

Volcanoes and hot springs are abundant in Iceland (Fridleifsson, in Cataldi *et al.*, 1999), having been exploited since the first settlement in the 9th century. The utilization of hot water was not only seen in bathing sites, but also as a heating source for homes and farms. The utilization of by-products such as sulfur and salt were recorded in official documents from the 13th century, describing exportation rights of Icelandic sulfur to Norway.

In Russia and former USSR countries (Svalova in Cataldi *et al.*, 1999) both in Europe and Asia, old bathhouses were built and used, due to the influence of the eastern Roman Empire through the Black Sea region (see, for instance, Tiblisi, Armenia). But it was in Kamtchatcka and the Kuril islands, where hot springs, volcanoes, and geysers are abundant, that geothermal energy utilization was developed by inhabitants.

2. The Birth of the Geothermal Industry in Italy: Larderello

Let's go back to Italy, in Tuscany, where the modern geothermal age started (Burgassi in Cataldi *et al.*, 1999). In the second half of the 18th century, the Grand Duke of Tuscany made an inventory of all the natural resources of its country. In 1777, the water from natural manifestation in the lagoon (*lagoni*) of Monterondondo was analyzed and boric acid was found. In 1818, a private French company (owned by de Larderel, among others) started the exploitation of the chemical contents of the geothermal water, via evaporation and accumulation. The great innovation was the utilization of steam for boiling water, instead of wood: the first modern industrial application of geothermal heat started in 1827. Along with the utilization of the geothermal water, drilling technology was also improved (first geothermal well in 1832), from a few meters down to 200 m in 1870 (see Figure 1).

At the beginning of the 1900s, the chemical factory in Larderello produced high-quality pure boric acid, ammonium sulfate, and borax.

In 1904, the first experiment for the production of electricity, by the factory owner Ginori Conti took place, using a piston engine, coupled with a 10 KW dynamo, powered by clean steam produced through a heat exchanger: the first experiment was an "indirect cycle" as in the modern binary plants.

It was in 1913 that in Larderello the first power plant was commissioned: it had a production capacity of 250 kW, immediately followed by two 3.5 MW units, making use of the same "indirect cycle" concept. The first direct steam cycle was the Serrazzano unit, in 1923 with a production capacity of 23 kW, proving the feasibility of the commercial operation of the geothermal steam inside the turbines.

1775 - 1904	1904 - 1905	1913
Chemical production	Start of geothermal power production first experiments	First power plant (250kW.)

Figure 1   Historical photo of Larderello geothermal development.

In 1930, the total number of units in the Boraciferous region added up to 12,150 kW and 7.2 MW indirect and 4.9 MW direct cycles.

In 1939, a large plant, called Larderello 2, with six units of 10 MW each, was built, with additional units in Castelnuovo, Sasso Pisano and Serazzano, reaching the total of 132 MW (107 MW indirect cycle).

The Second World War destroyed all the industrial installation in the region, so the first stage of the modern utilization of geothermal energy ended tragically in 1944. The recent development after the Second World War will be followed in the country-specific chapters in this book.

3. The Geothermal Energy in Europe

Geothermal will be a key energy source in the European low-carbon energy mix. This source has unique features of not only being renewable and local but also continuously available. Europe has pioneered the exploitation of geothermal resources for a century and the European Union (EU) still maintains a leading role due to the development of new advanced technologies; geothermal is a unique energy source that can provide a significant share of electricity, heating and cooling, in 2020 and beyond. It is indeed renewable, local and continuously available, as it is not dependent on climate conditions. Europe has pioneered the exploitation of geothermal resources for a century and the EU still maintains a leading role due to the development of binary cycle for low-temperature resources and of enhanced geothermal systems (EGS), an already demonstrated breakthrough technology, allowing the production of geothermal power as well as heating and cooling everywhere.

3.1. Geothermal Market Development

Over the last five years, we have witnessed a resurgence of interest in geothermal power, after nearly a decade of only small development in capacity in the deep geothermal sector, both for electricity and for direct uses (mainly district heating).

A substantial number of projects have been developed throughout Europe, and geothermal energy is on its way to becoming a key player in the European energy market.

Geothermal District Heating continues to show dynamic development in Europe, with 10 new systems (eight in France, one in Italy, and one in the Netherlands), having an installed capacity of some 100 MWth, becoming operational in 2015, and more than 200 projects now under development. The total installed capacity from the 256 plants in Europe is now 4.6 GW_{th}. Of this, 176 plants are located in the EU, producing around 4300 GWhth (EGEC Market Report, 2016).

The power sector also continued to grow, but at a much slower pace than is possible. Twelve new plants were commissioned between 2014 and December 2015, 10 of which are in Turkey, one in Bavaria, Germany, and one in Tuscany, Italy. Overall, the newly installed capacity amounts to 266 MWe. There are now 84 power plants in Europe, representing a total installed capacity of nearly 2.2 GWe. Fifty-two of these plants, with a total installed capacity of 991 MWe, are in the EU-28, producing roughly 5.9 TWhe annually. The fast development of the Turkish market and the 34 projects currently being developed will be key in the expected growth to ca. 2.7 GWe by the end of this decade.

The SET Plan integrated roadmap published in October 2014 has included the geothermal potential in 2030 and 2050: The potential in 2050 is stated as 320 GWe installed capacity worldwide, with about 90–100 GWe in Europe. Considering the high load factor of geothermal power plants with an average of about 70–80%, geothermal energy could provide 550–700 TWh of electricity per year. The short-term targets of this challenge should be to achieve an installed geothermal power capacity in Europe of >3 GWe (of which 1.5 GWe in EU27) by 2020, and aiming at >10 GWe by 2030.

Geothermal power can contribute to reduction in the use of fossil fuels, foster local development, decarbonize our electricity sector, and diversify Europe's energy mix. It can also provide protection against volatile and rising electricity prices. Contrary to some other types of base-load power plants, geothermal plants are also flexible and reliable, that is, available under any circumstances. Thanks to these features, geothermal technologies could therefore positively contribute to decarbonization and other EU objectives. Research, development, and innovation will be much needed to develop the new generation of flexible renewable energy technologies as well as to improve the flexibility of their electricity production.

The total installed capacity from worldwide geothermal power plants is 12,635 MWe, with 73,549 GWh of electricity produced. The forecast is to have 21,443 MWe installed by 2020. To reach the predicted figure by 2020, based on an accurate accounting of all the existing projects at an executive stage, a clear change in the present linear growth trend is necessary (see Figure 2).

An increase of about 1.7 GWe in the five-year term 2010–2015 has been achieved (about 16%), following the rough standard linear trend of approximately 350 MW/year, with an evident increment of the average value of about 200 MWe/year in the precedent 2000–2005 period (Bertani, 2016).

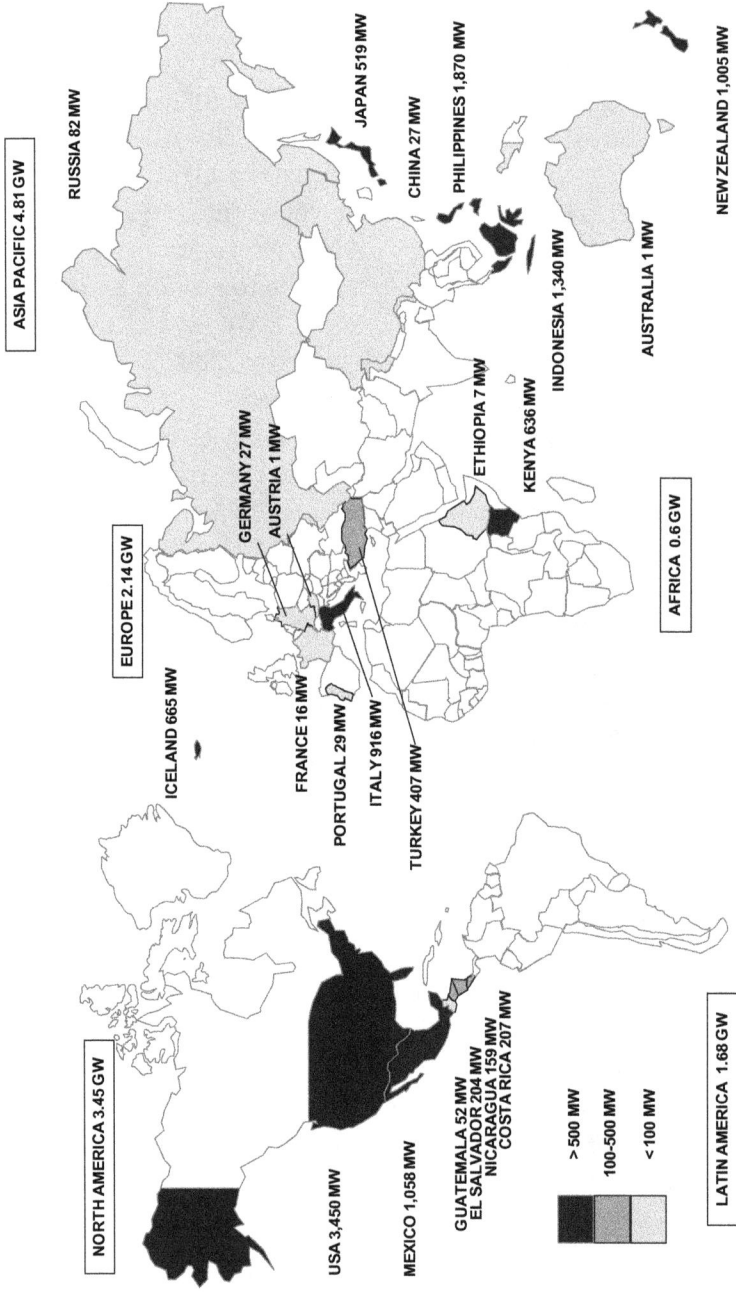

Figure 2   World geothermal installed capacity.

4. Presentation of this volume

This book will represent a snapshot of the present situation of geothermal Europe; after a general overview by Philippe Dumas (Geothermal Energy in Europe), we have decided to present the most suitable technology for continental Europe's electricity generation: binary plants (Joseph Bonafin) and the most promising technique for heat utilization, heat pumps (Ladislaus Rybach and Burkhard Sanner).

The most important geothermal countries have been highlighted in chapters dedicated to each country:

- Iceland (Ólafur G. Flóvenz and Brynja Jónsdóttir)
- Italy (Adele Manzella, Assunta Donato, Gianluca Gola, Alessandro Santilano and Eugenio Trumpy)
- Turkey (Sakir Simsek)
- The Netherlands (Jan-Diederik van Wees, Maarten Pluymaekers, Damien Bonté, Serge van Gessel, and Hans Veldkamp)

As a conclusion, a discussion on the important aspects of policy and regulation in Europe is presented by Luca Angelino.

I would like to thank all the authors for their valuable help and high level of contribution to the volume.

References

Bertani R., 2016. Geothermal power generation in the world 2010–2014 update report, *Geothermics*, 60, pp. 31–43.

Cataldi R., Hodgson S.F., and Lund J.W., editors, 1999. *Stories from a Heated Earth; Our Geothermal Heritage*. Geothermal Resource Council, International Geothermal Association, Sacramento, CA.

EGEC, 2016. "European Market Report 2015".

Chapter 1

Geothermal Energy in Europe

Philippe Dumas

European Geothermal Energy Council
2 Place du Champ de Mars, B-1050 Brussels
p.dumas@egec.org

Geothermal development in Europe dates back to more than a century, but the market is still at the infancy stage. We have witnessed within the last five years a resurgence of interest in geothermal power, after nearly a decade of only small development in capacity in the deep geothermal sector both for electricity and for direct uses (mainly district heating). A substantial number of projects have been developed throughout Europe, and geothermal energy is on its way to become a key player in the European energy market.

The current market conditions do not allow this development; many nontechnical barriers still need to be removed. A new generation of geothermal technologies is also needed for answering the challenges of the next decade for the European energy system. If the energy transition has to be successful, we have to think about an optimal scenario in terms of costs and affordability for the customers and the citizens.

Geothermal will be a key enabler, a local and stable source of renewable energy, and its role will be crucial in the future energy system.

1.1. Introduction

Geothermal is a unique source of energy. Geothermal energy and Europe have a long history of developing together. In Europe, geothermal energy is the source of both electrical power for a century and heating and cooling for hundreds of years. It is one of the renewable energy sources. But both its capacities and its development in Europe make it quite unique. Geothermal is able to answer the challenges for the future European energy market design being capable to produce as base load electricity production and as a flexible power generation. Geothermal is also able to provide heating and cooling solutions for the building sector and the industry. Its past and current developments make this technology rather unique. It has typical barrier, notably economical such as its very high capital intensity and its risk profile; and legal such as the regulatory procedures for underground development in the different European countries.

Geothermal is officially defined, since 2009, by the institutions of the European Union (EU), in the Directive on Renewable Energy Sources (2009/28/EC) as the following: "geothermal is the energy in form of heat beneath the surface of the earth".

Geothermal energy is the heat from the Earth, or, more precisely, that part of the Earth's heat that can be recovered and exploited by man. Evidence of terrestrial heat is given by volcanoes, hot springs, and other thermal manifestations. Earth's temperature is increasing with depth, under a gradient of 2–3°C/100 m. The total heat flux from the Earth's interior amounts to ca. 80 mW_{th}/m^2. It provides us with an abundant, renewable, almost infinite source of energy.

However, only a fraction of it can be utilized by people. Our utilization of this energy has been limited to areas in which geological conditions allow a fluid (liquid water or steam) to "transfer" the heat from deep hot zones to, or near, the surface, thus giving rise to geothermal resources.

In Europe, as detailed in the following chapters, the potential to develop geothermal is huge. It was first exploited during ancient times, by the Greeks and the Romans, for its capacity to produce heating and cooling for recreational purposes such as spas and bathing. Until the 12th century, geothermal energy was indeed mainly exploited as direct uses for leisure and sometimes for the agro-food production. Power production and heating and cooling applications in the buildings sectors from geothermal started during the last century.

As presented in the map (Figure 1.1), geothermal energy can be developed all over Europe. The regions with high enthalpy (in red) are eligible for power production. Areas with medium enthalpy (in orange and dark yellow) can be used for the installation of power

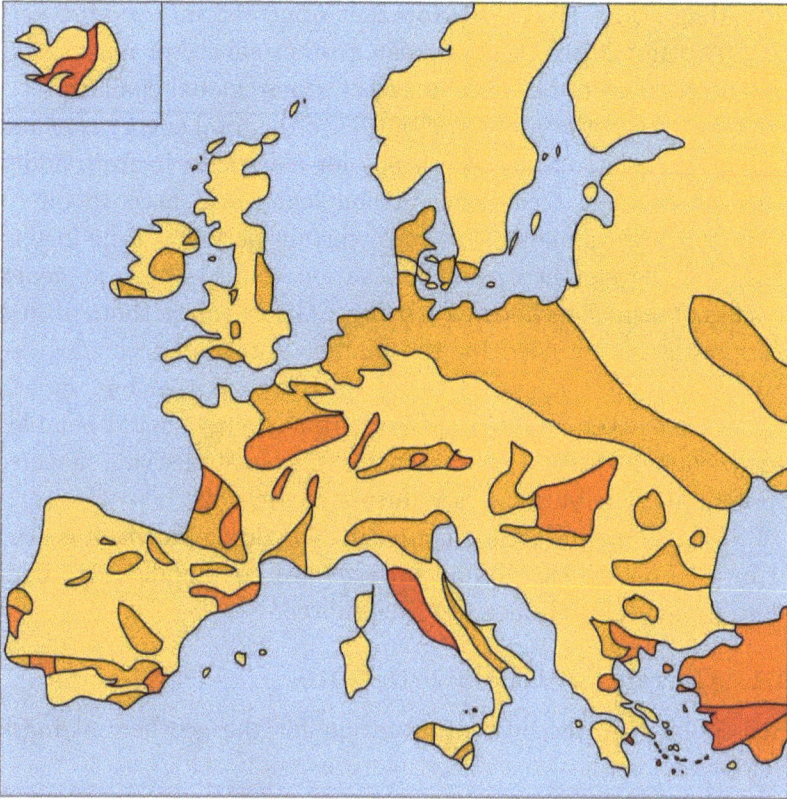

Figure 1.1   Map of geothermal energy in Europe (EGEC).

plants running at low temperature (binary plants) and of direct uses applications such as district heating (DH) systems. Finally, the rest of Europe (light yellow) is eligible for a power generation with engineered/EGS and for a heating and cooling production assisted with heat pumps. All these areas can also be exploited for underground thermal energy storage (UTES).

The fight against climate change and its translation into international agreements within the United Nations Framework Convention on Climate Change (UNFCCC) require Europe to decarbonize its economy. The energy transition toward a low or a zero carbon energy has already started. The next five to 10 years will be crucial for designing the future energy mix if we wish to answer the 2050 climate and energy challenges. It means that the current period is key for deciding about future development of geothermal production in 2020, 2030, and 2050. The European geothermal sector must demonstrate its abilities to be a key answer to these global challenges.

Today, the geothermal contribution to the total energy consumed in Europe is relatively small, except for electricity in Italy and for heating in Sweden. Geothermal technologies still need to be supported (not always financially) for a certain period of time in terms of research, development, and innovation toward the next generation of geothermal technologies being EGS or smart thermal grids. A support is also needed for the market uptake of technologies at a precommercial stage or those already competitive but suffering from unfair market conditions. Geothermal energy should be able to develop enough by 2020 for convincing European decision makers of its potential contribution to the future energy mix.

Two of the main factors hampering the development of geothermal in Europe are that its advantages are not highlighted enough and still many barriers need to be removed.

1.1.1. Current Situation: 2015–2016

When looking at the current situation for the geothermal market development, we notice a mixed picture:

- The geothermal DH sector is developing.
- The situation for geothermal electricity has also developed quite well recently, especially in Turkey, with preparation times being longer than expected in the EU countries.
- For shallow geothermal, the development is not satisfactory at all, with several factors hindering the desired growth.

1.2. The Geothermal Power Production

There are 88 electrical power plants in Europe representing a total installed capacity of 2285 MWe and a production of around 12 TWh of electric power (Figure 1.2).

Geothermal plants in Europe are characterized by a high availability (amount of time that a plant is able to produce electricity

Gross electricity production (GWh) in selected countries in 2014

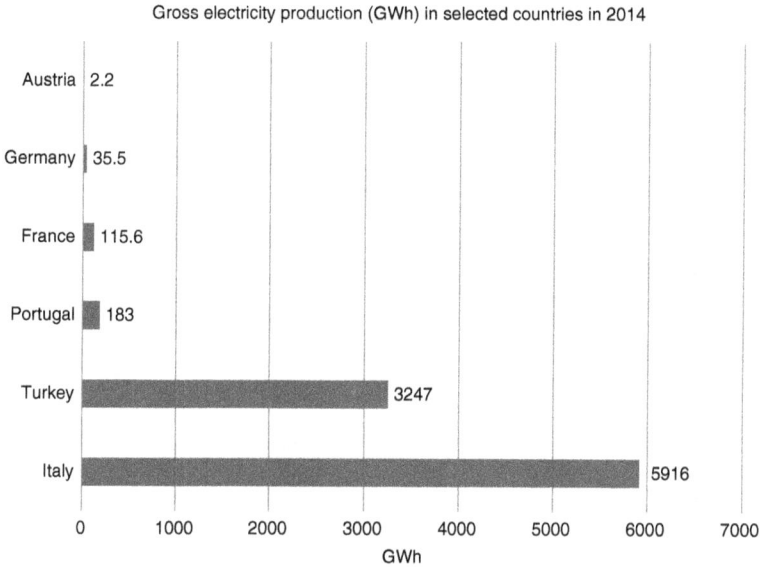

Figure 1.2 Graph of geothermal power production in Europe in 2014 (EGEC, 2015).

over a certain period, typically a year, divided by the amount of the time in the period, i.e., 8765–8766 h) and net capacity factor (the ratio of the actual output of the geothermal plant over a period of time, to its potential output if it were possible for it to operate at full nameplate capacity indefinitely), typically in excess of 80%. Some geothermal plants operate at 100%.

The average growth rate for the last five years is of 5%, with around five to ten new power plants inaugurated per year representing an annual new capacity installed of about 100 MWe (Figure 1.3).

It is important to highlight that more than 200 projects are currently under development or investigation, and it means that the number of plants operating in Europe could double in the near future. But the geothermal power market is not developing as quickly as expected. There are three main reasons for this:

- First, the vast geothermal potential is still underestimated and thereby there is an urgent need to increase awareness of its advantages especially for decision makers and investors.

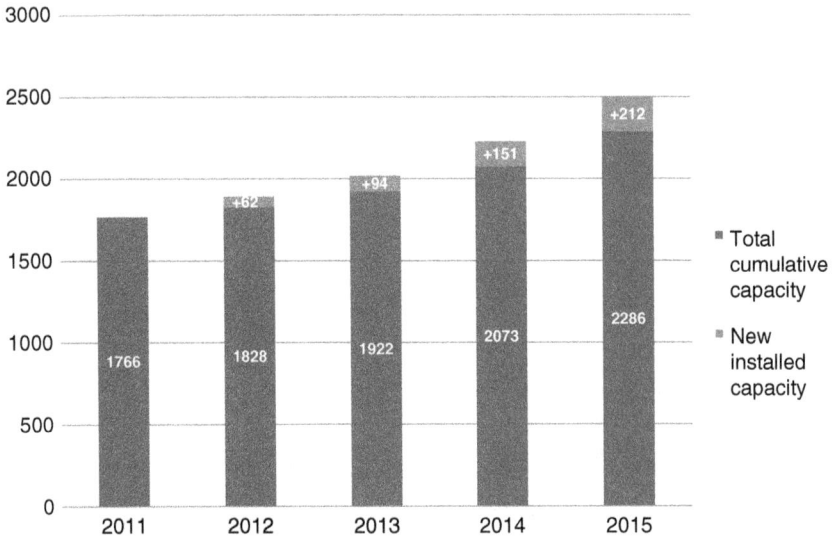

Figure 1.3   Graph of cumulated installed capacity in Europe 2011–2015 (EGEC, 2015).

- Second, much more financial support should be brought to the geothermal sector. Support schemes are crucial tools of public policy for geothermal to compensate for market failures and to allow the technology to progress along its learning curve. Funding allocated to geothermal energy is negligible compared to that which is allocated to other technologies.
- Finally, beyond exploration, the bankability of a geothermal project is threatened by the geological risk. Risk insurance funds for the geological risk already exist in some European countries (France, Germany, Iceland, the Netherlands, and Switzerland). The geological risk is a common issue all over Europe. Collaboration between member states is desirable; it can allow them to save money and trigger the uptake of a valuable technology alike.

The main geothermal markets for electrical generation in Europe are Turkey, Italy, Germany, and France. Today, Italy continues to dominate the market but Turkey is the hottest market and more than 50% of the projects under development representing a capacity of around 650 MWe by 2017 are planned in this country. France and

Germany already have a few geothermal power plants in operation and many permits have been requested and granted. These countries can now be seen as future hot markets, especially regarding low-temperature production and EGS.

Kizildere geothermal field was the first field utilized in 1974 for electricity production in Turkey. The Kizildere pilot plant had a capacity of 0.5 MWe. Larger-sized power production in Kizildere started in 1984 with increased installed capacity of 15 MWe. New geothermal power plants have been installed since 2006 with a real boom between 2009 and 2015. Finally, it is worth mentioning that a significant number of licenses have been granted by the Energy Market Regulatory Authority (EMRA) of Turkey. The total installed capacity may further grow until around 1 GWe by 2020. Turkey is therefore set to overtake Italy as the leading European geothermal country in terms of installed capacity by the end of the decade. After the liberalization of the Turkish electric market, many players are now conducting exploration activities, preparing the basis for the future development of geothermal energy in Turkey.

It can be noted that all projects are developed by Turkish project developers; some are in consortium with a foreign companies (co-project developer, equipment providers, utilities, turbines providers etc.).

Italy still represents more than 50% of the European capacity with around 915 MWe installed capacity today in Tuscany. Some new plants are currently being developed and should become operational by 2020.

A Legislative Decree (No. 22 of February 11, 2010) has liberalized access to the geothermal market, allowing new players to enter into the geothermal business by applying to the regional authority for an exploration lease. In recent years, several new leases have been requested in many different Italian regions.

More than 130 requests for research permits regarding geothermal exploration and exploitation have been presented in the last couple of years. Of these, 50 have been granted, 37 of which are in the Tuscany region, and around 40 under evaluation.

France is the real new hot market in Europe with a particular focus on EGS. Geothermal electricity has been produced in France since 1986 at the Bouillante 1 plant in Guadeloupe (double flash of 4 MWe). This plant has been extended by 11 MWe with Bouillante 2 inaugurated in 2005. The total capacity is currently 14.7 MWe and the production is about 80 GWh. Plant extension with Bouillante 3 is under development. Most activities today are in Alsace. The Soultz-Sous-Forêts power plant has a net capacity of 1.5 MWe for net production, after 25 years of research and about €80 million of investment.

The first EGS plant has been producing electricity since 2008 and is connected to the national grid, with four wells at a depth of 3500–5000 m. The next stage of development will focus on equipment efficiency.

The recent inauguration of the ECOGI project in Rittershofen, an EGS to heat a factory has proven the potential of the whole region north of Strasbourg.

More than 10 licenses are granted in France, and the same number of licenses has been requested in recent years and is currently under examination.

Germany has been seen as an important new entrant for some years. Power production started in 2003 with the Neustadt-Glewe geothermal plant, which now operates only as heat plant (turbine dismantled in 2012). With the inauguration of new geothermal plants between 2013 and 2015, there are now nine plants in operation representing a capacity of 36 MWe.

Several geothermal power projects are expected to be commissioned in the next years. A total of 15 projects are under development most of which are concentrated in Bavaria and in the Upper Rhine Graben area.

1.2.1. The Geothermal Heating and Cooling Production

Geothermal DH continues to show dynamic development in Europe, with 23 new systems (nine in France, two in Italy, three in the Netherlands, five in Hungary, three in Germany, and one in Austria), having an installed capacity of some 100 MW$_{th}$, becoming operational in 2015, and more than 200 projects now under development (Figure 1.4). The total installed capacity from the 257 plants in Europe is now 4.7 GW$_{th}$. Of these, 177 plants are located in the EU, producing around 4300 GWh.

DH is the geothermal sector currently with the most dynamic development and the most interesting perspective in the coming years. The renewed momentum since 2009 continues, with new countries installing geothermal DH systems in the past year. The technology is developing: smaller systems, targeting shallower resources, and assisted by large heat pump systems have been installed. In France, more triplet systems have been installed.

There are 257 Geothermal DH plants (including co-generation systems) in Europe representing a total installed capacity of more than 4.7 GW$_{th}$ and a production of some 13.1 TWh. The main GeoDH markets are in France (52 systems), Iceland (32), Germany (23), and Hungary (24).

Cumulative installed capacity in Europe 2011–2015 (MW$_{th}$)

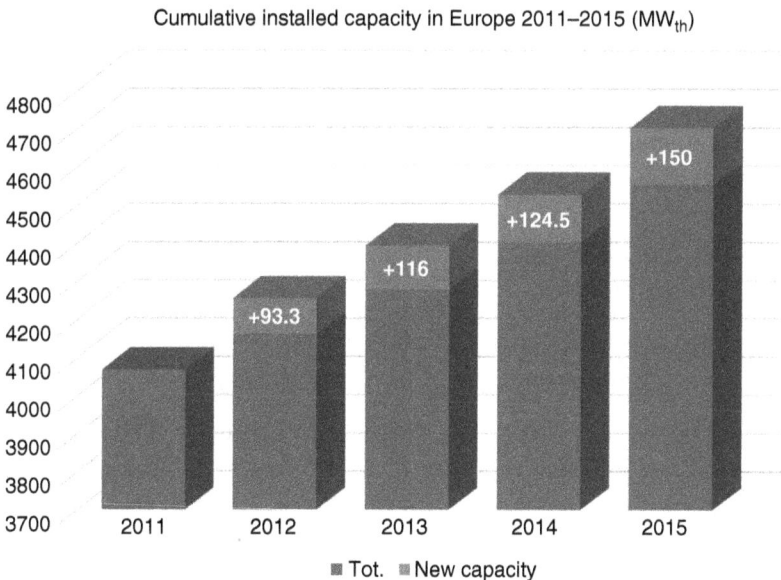

Figure 1.4   Graph of cumulated installed capacity in Europe 2011–2015 (EGEC, 2015).

The hot markets are also mainly in Germany (43 new systems being developed or upgraded), France (46), and Hungary (17). It is of interest to highlight the situation in Hungary, a country with a long tradition in geothermal DH, which now sees new development.

One important new actor in the direct use/GeoDH market is the Netherlands where eight deep geothermal systems for heating and cooling have been installed recently, and where three more are planned.

CHP helps geothermal to become more economically attractive by recovering waste heat for heating and cooling purposes. Until now, only a few combined heat and power geothermal plants supplied DH systems, but this situation is rapidly changing. As a matter of fact, EGS (CHP) provides more opportunities for GeoDH systems.

In conclusion, it can be stated that 26 European countries (21 of which are EU member states) show deep geothermal activity, proving that geothermal can be developed almost anywhere in Europe.

For shallow geothermal energy (ground source heat pumps [GSHP] and UTES), the overall installation growth is continuing at a steady pace. This brought installed capacity up to at least 19,000 MW_{th} by 2014, distributed over about 1.4 Mio GSHP installations. The countries with the highest amount of geothermal heat pumps are Sweden, Germany, France, Switzerland, and Norway. These five countries alone account for ca. 69% of all installed capacity for shallow geothermal energy in Europe. An interesting development can be stated for the countries in Central and Eastern Europe. While the absolute numbers in these countries still are low, there is a quite positive development.

1.2.2. Its Characteristics: A SWOT Analysis

A strengths, weaknesses, opportunities, and threats (SWOT) framework is a planning and evaluative tool. It intends to illustrate existing capacities and future possibilities. In the case of the European geothermal sector, the SWOT will cover power and heat technologies and it will treat the issues that the customers and citizens, decision makers, project developers, and financiers must confront.

Geothermal strengths are numerous:

- It is a unique source of renewable energy, producing electrical power, heating and cooling, and domestic hot water. It can then provide energy to all the sectors. Its principal strength is that it is both a base load and a dispatchable power and heat source, available 24 h per day, produced locally.
- Geothermal energy in Europe has been developed for centuries and it is currently developing a new generation of geothermal technologies such as EGS and smart thermal grids.
- It can generate power for helping to stabilize the grid, to make it more flexible and to ensure security and firmness of power supply.
- It produces heating, cooling, and domestic hot water for buildings (small and large, for renovation and new construction) and for the industry (mainly low and medium temperature), being also able to offer UTES solutions.
- It has the advantage of presenting all costs to the consumers, as little systems costs and externalities have to be integrated. It participates in the development of the local economy creating jobs.
- The sector is still rather small but with a lot of experience, which can be exported to other continents.

Its weaknesses are the following:

- Upfront costs of project development are very high, with another order of magnitude compared to other Renewable Energy Sources. It is capital intensive so it needs to be developed by rather large companies and supported by large financial institutions.
- Its risk component constitutes also one of the main weaknesses.
- In terms of development and marketing, it is a relatively small sector compared to others, and it is complex to promote due to two main factors: it has no visual impact, and it is technically harder to understand than other energy technologies. Moreover, the potential is not defined enough and the resource characteristics are not well mapped and identified. Some regulatory barriers create uncertainty on the future development of the sector.

But many opportunities exist for developing geothermal in Europe. It is one clean energy solution to decarbonize the European economy, fight climate change, and implement the COP21 Paris agreement signed last December 2015. The European power sector is living a revolution with an internal market, more flexibility, and a decentralized approach. The current reform of the European electricity market design should offer new opportunities as the demand for dispatchable power and security of supply is growing. The sector should also be able to lower its costs and its uncertainties to develop new markets in Europe.

The global market is also an opportunity for the European geothermal sector as new developments on power and heat can be seen in Africa, Asia/Oceania, and South America.

Indeed, the heating and cooling sector can expand too. New trends in the DH sector with low-temperature and smaller systems offer interesting developments for geothermal. The potential of geothermal in the agro-food industry and other direct uses in the industry and services sector is also largely untapped. Finally, more attention should be paid on UTES and geothermal heat pumps for construction or near-zero-energy-building and for buildings renovation.

The main threats the geothermal sector is facing is the persistence of unfair market conditions and the lack of recognition of the role that geothermal can play in the energy mix.

For the power sector, it means that the potential is not realized and the contribution in terms of generation remains marginal, and EGS will not be developed.

For heating and cooling, the geothermal sector stays a niche market, and many countries will not develop geothermal.

But these challenges can be solved by the sector and all barriers highlighted above should be removed when achieving the European energy internal market.

1.3. Trends

When presenting the development of an energy technology, we should all try to assess the global picture having in mind the long-term perspective, and not just the three to five last years. This relative picture helps to have the right assessment. Regarding geothermal, the first power plants date back to 1913, but the development was really low until the 1970s. The growth during the last 30 years, although it is lower than for some other RES, has to be highlighted. When looking at the current development, we can be sure that it is just the start of a much larger development.

1.3.1. Electricity Development from 1913 to 2016

The first experiment to produce geothermal power was done in Italy in 1904 by Prince Ginori Conti (Figure 1.5).

In 1818, François Larderel (French) started the first industrial application: use steam to run pumps. He gave his name to the village of Larderello which housed the workers.

Piero Ginori Conti married in 1894 Adriana de Larderel (cousin) and became director of the family company. Prince Piero Ginori Conti tested the first geothermal power generator on July 4, 1904, at the Larderello dry steam field. It was a small generator that lit four light bulbs, before a first 20 kW plant.

In 1913, the first geothermal electricity plant started its operation in Larderello (Italy) and was composed of an alternator Ganz 250

Figure 1.5   Photo of Prince Ginori Conti (courtesy of ENEL).

kW, three-phase 50 Hz, with a voltage of 4500 V coupled with a Tosi Parsons turbine of 350 hp.

Until the 1970s, only a few other plants were installed in Europe, that is in Italy and Iceland, followed by France (Guadalupe) and Turkey. And all of them were high temperature with flash/dry steam turbine technology. With the development of new turbine technologies, that is, binary cycle (Kalina and Organic Rankine Cycle [ORC]) for low and medium temperature, more countries (Austria, Germany, Portugal) have installed geothermal power plants in the last 40 years (Figure 1.6).

Currently there are 88 geothermal power plants in six European countries (Italy, Iceland, Turkey, Portugal, France, Germany, and Austria) for a total installed capacity now amounting to around 2.3 GWe, producing some 12 TWh of electric power every year (EGEC Market Report 2015).

According to the EGEC Geothermal Market Report 2015, there are 32 projects currently under development in Europe. In addition, 176 projects are now being explored.

Italy dominates the market with more than 50% of the European capacity, i.e., 915 MWe. After the liberalization of the Italian

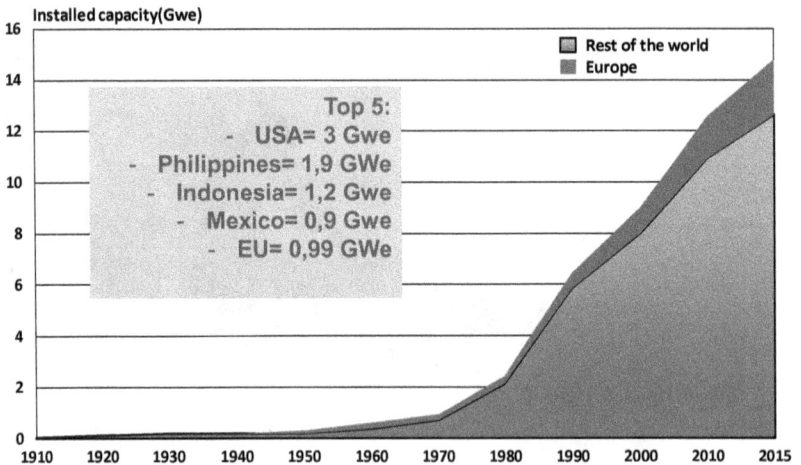

Figure 1.6 Development of the geothermal power capacity globally and in Europe 1913–2015 (EGEC).

geothermal market (legislative decree No. 22, 2010), more than 130 applications for research permits for geothermal exploitation and development have been submitted. Many new players are now operating in exploration activities, preparing for the future development of geothermal energy in the country.

Iceland has installed eight power plants representing a capacity of 662 MWe. Nearly 90 MWe are currently being developed. In addition, five more projects are being investigated, notably one, the Iceland deep drilling project, which could provide a very large amount of electricity if successful in exploiting supercritical resources.

In Turkey, the market is booming. According to the projects under development and investigation, the installed capacity should grow from 632 MWe today (28 plants) to double and to reach 1 GWe in 2020. Turkey continues today to be the hottest market in terms of newly installed capacity this year and more than 50% of the projects under development representing a capacity of around 700 MWe by 2017. With more than 10 projects under development and 13 under investigation, the total installed capacity may further grow until around 1 GWe by 2020.

France (Guadeloupe) and Portugal (Açores) have been developing geothermal electricity power plants on Atlantic islands since the 1980s; this development is continuing with the Geothermie Bouillante 3rd unit and the Pico Vermelho plant for 2016. France is the home of the first EGS pilot project (Soultz), which was inaugurated in 2008. At least 12 other EGS projects are being investigated with more and more permits for research and exploration being awarded by the government.

1.3.2. Power Potential: 2030–2050

The Geoelec project (www.geoelec.eu), supported by the EU through the Intelligent Energy Europe program, and coordinated by EGEC, has created a European map showing an overview of the location of geothermal resources, which can be developed in 2020, 2030, and 2050. The map is based on a unified reporting protocol and resource classification for geothermal resource assessment.

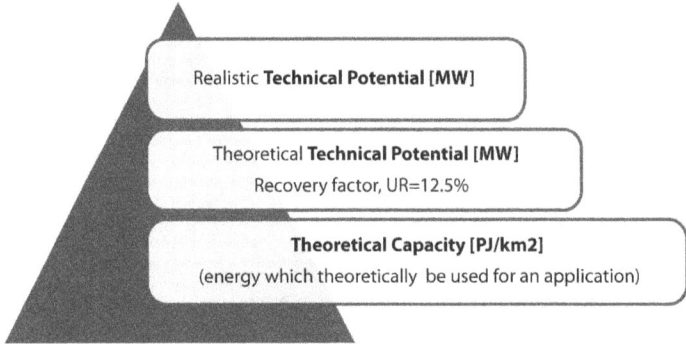

Figure 1.7   Schematic workflow from theoretical potential to realistic potential. *Source*: *Geoelec Projeet* (2014).

The resource assessment of the geothermal potential for electricity generation is the product of the integration of existing data provided by the EU-28 countries and a newly defined methodology building on Canadian, Australian, and American methodology. The objective is to present the economical potential of the geothermal power production in Europe in 2020, 2030, and 2050 (Figure 1.7).

The geological potential (heat in place) is translated to an economical potential using a levelized cost of energy (LCoE) value of less than €150/MWh for the 2030 scenario and less than €100/MWh for the 2050 scenario. These two figures are considered to be the grid parity at the respective times with a full internalization of the system costs and other externalities.

- The production of geothermal electricity in the EU in 2013 is 6 TWh.
- The NREAPs forecast a 2020 production in the EU-28 of ca. 11 TWh.

The economic potential for geothermal power in 2030 has been calculated using a LCoE value of less than €150/MWe:

- 34 TWh for the EU-28.
- 174 TWh for the total potential in Europe.

The trend will be an acceleration of the growth rate already 2014 and continuing until 2030. We can notice in the following map:

- Already geothermal power is very competitive in Iceland, Italy, and Turkey today.
- By 2030, several regions will be able to produce geothermal electricity at low costs, i.e., the Pannonian Basin, the Rhine Graben area, and some regions of Spain, UK, France, the Netherlands, Germany, amongst others.
- In the rest of Europe, pilot EGS projects should be developed in order to bring down costs in these regions and to allow the technology to progress along its development curve.

The economic potential for geothermal power in 2050 using a LCoE value of less than €100/MWh:

- 2570 TWh for the EU-28.
- Ca. 4000 TWh for the total potential in Europe.

The potential is enormous, especially when EGS becomes very competitive after 2030 (Figure 1.8, Geoelec final report):

- Geothermal power could be produced competitively all over Europe, in certain regions of each country in the EU.
- In other regions, although the drilling costs could remain important, the development of combined heat and power projects will make them competitive economically.

Geothermal power production has a substantial potential in Europe. Technology does not only include the geothermal steam plants in Iceland, Italy, and Turkey, but also the use of hot water in binary systems (currently applied in the previous counties and Austria, Germany, France, and Portugal). The next generation geothermal technology, EGS, aims at opening the geological limits for current geothermal power sites and strives to make geothermal power an option almost anywhere in Europe. Most geothermal power plants are capable (or can be made capable, by tuning the flow rate of the in-well production pumps) of responding to command from system operators to ramp output up and down on demand

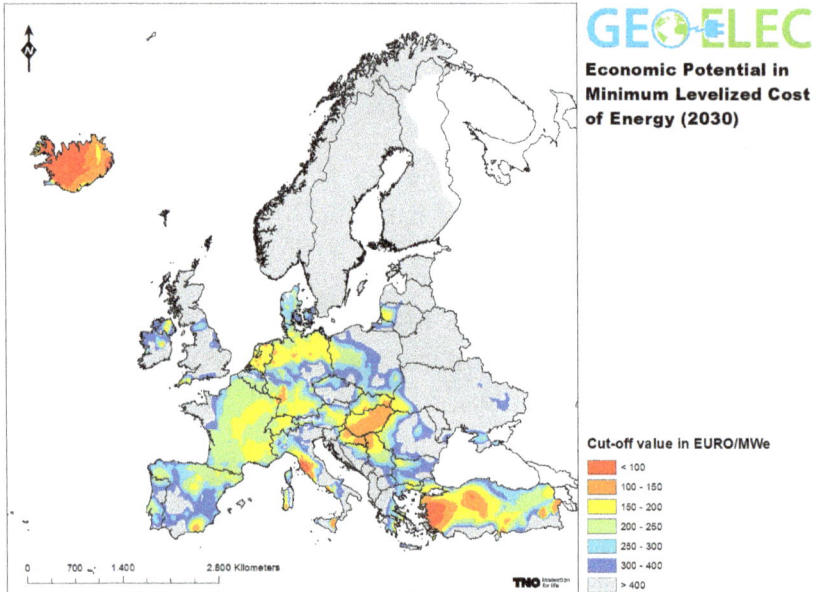

GE ELEC

Economic Potential in Minimum Levelized Cost of Energy (2030)

Cut-off value in EURO/MWe

▮	< 100
▮	100 - 150
▮	150 - 200
▮	200 - 250
▮	250 - 300
▮	300 - 400
▮	> 400

0   700   1.400   2.800 Kilometers

Figure 1.8   Economic potential of geothermal power production in Europe in 2030.
Source: *Geoelec Project* (2014).

and thereby can provide the necessary flexibility to the electricity system. Geothermal power can be of help to improve grid flexibility, for example, by installing geothermal plants to ensure a regional approach, a step between centralized and decentralized systems; or have a system approach with regional/local security of supply, and so mitigate infrastructure and storage costs.

- Make full use of all geothermal power options by lifting EGS technology from the level of single demonstration plants to that of a widely deployable and competitive technology, and open ways for using all the potential, from low temperature to unconventional geothermal resources. The short-term targets of this challenge should be to achieve an installed geothermal power capacity in Europe of >3 GW (of which 1.5 GW in EU27) by 2020, and aiming at >10 GW by 2030.
- Keep production cost for electricity from geothermal resources low, by decreasing installation and operation cost of the power

plants, by increasing longevity of installations, by optimizing efficiency and power output. A substantial part of installation cost is drilling, and R&D here needs to focus on new drilling technologies and concepts. Also further development in power conversion technology at lower temperatures is required in order to enlarge the area of application and increase efficiency (Figure 1.9, EGEC, *Strategic Research Priorities for Geothermal Technologies* (2012) EGEC. *Geothermal Technology Roadmap* (2014)).

- Include geothermal power in grid optimization schemes, and use its advantages as a base load, flexible, sizable, controllable, and local resource (one of the few among renewables). Enable geothermal CHP to help in the frame of economical local development (cascade uses), or in the combination with smart electrical and thermal grids, for the latter including underground thermal storage.

- Enhance the current use of geothermal energy for heating and cooling by reducing cost of exploration, drilling, and installation by improving longevity of material and efficiency of operation, and by opening new areas for heat from deep geothermal resources through EGS technology (cf. issue 1 above). In particular with geothermal heating and cooling, nontechnical issues like investor

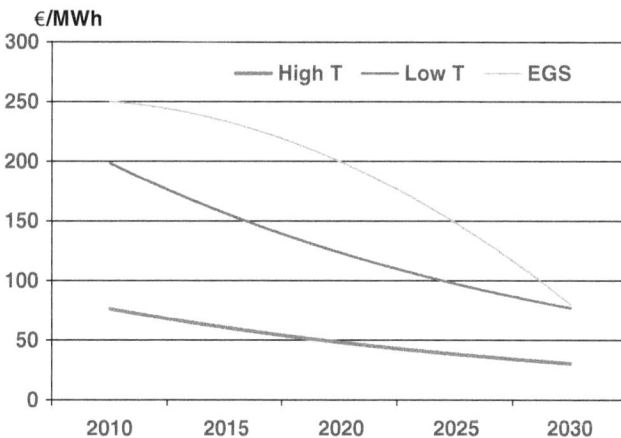

Figure 1.9   Potential cost reduction of geothermal power technologies by 2030.
Source: Geothermal Technology Roadmap (2014).

awareness, city planning, regulations concerning the resource, etc. can create substantial barriers against further deployment, and thus need to be addressed in priority. Also activities outside the realm of geothermal energy *sensu strictu*, like development in thermal storage and in (smart) thermal grids, are crucial to make full use of the geothermal heat resources.

1.3.3. Geothermal Heating and Cooling: Until 2015 to Post 2015

The heating and cooling sector represents 50% of our energy demand, and geothermal is becoming more and more attractive as a competitive renewable heating source, as there is a dual need to decarbonize this sector, while securing the provision for heating at an affordable price for consumers.

To cover demand for heating and cooling, geothermal energy offers vast resources with numerous technical options, ranging from supply to individual buildings using geothermal heat pumps to providing heat (and cold) for whole cities or city quarters through large DH networks.

1.3.3.1. Geothermal DH and direct uses

The "hot" GeoDH main markets in Europe are located in France (the Paris basin, and renewed activity in the Aquitaine basin), Germany (the Molasse basin, Munich), and Hungary, but it is important to emphasize that geothermal DH systems can be installed in all European countries. We have recently seen important new markets emerging, especially in the Netherlands and the UK. By 2020, almost every European country will use GeoDH.

The first regions to install GeoDH were those with the best hydrothermal potential, however with new technologies and systems, there is an ever increasing batch of regions that are developing geothermal technology for heating and cooling applications. Systems can be small (from 0.5 to 2 MW_{th}), or large, with a capacity of 50 MW_{th}. There are some new DH schemes that utilize shallow geothermal resources, assisted by large heat pumps. In addition, it is worth mentioning that there are different management schemes in place according to the size of the systems, some DH systems are developed and managed by local authorities while larger systems, with higher capital costs, are developed by utility companies. The first geothermal DH plants have been installed in these countries for many years.

For example, in Iceland, in November 1930, the first geothermal DH system in Reykjavik was installed via a 3 km long pipe. Over the following decades, geothermal DH became widespread, as a result of local initiatives. After the oil shocks of 1973 and 1974, GeoDH became even more important for Iceland.

The larger scale demands of energy production, as well as advances in drilling and equipment meant deeper wells, higher temperatures, and more sophisticated technology. Currently, well over 90% of Icelandic homes are heated by geothermal energy, the highest percentage in the world. Most of the DH in Iceland comes from three main geothermal power plants, producing over 800 MW_{th}:

- Svartsengi combined heat and power plant (CHP).
- Nesjavellir CHP plant.
- Hellisheidi CHP plant.

In France, the first GeoDH systems were installed in the 1960s in the Paris basin, which pioneered the doublet well concept of heat farming, which was further replicated in the Paris basin GeoDH undertakings, initiating the typical doublet well array and exploiting the deep-seated Dogger carbonate reservoir. Between 1982 and 1986, more than 70 plants were installed in France.

The reduction of the oil and gas prices in the 1980s, combined with technical shortcomings (mainly scaling and corrosion) stopped this growth for about 20 years.

Today in France, there are 52 GeoDH systems in operation and 46 others are being developed.

GeoDH in Hungary dates back to 1960, in the city of Szentes, and in Germany, the first geothermal doublet for DH was completed in 1984 in Waren/Müritz. Now we can see in Hungary and Germany the same story as in France: a renewal of momentum from 2008 on.

Based on Europe's geothermal potential, geothermal energy could contribute much more significantly to the decarbonization of the DH sector. A considerable expansion of the DH sector is expected in the EU28 until 2050; indeed, geothermal heat through future DH systems could be available for 26% of the population. It is crucial to target areas with urban density to ensure the economic sustainability of the project.

Around 20% of the EU population are located in regions where the temperature at 2000 m deep is higher than 60°C, so are directly suitable for geothermal heating and cooling.

The GeoDH project EGEC, *Geodh Final Report* (2014) supported by the EU through the Intelligent Energy Europe program and coordinated by EGEC aims, among other things, to provide an interactive web-map viewer that shows areas with good geothermal potential for DH and heat demand.

The web-map indicates the existing DH systems, including GeoDH systems, in Europe. Moreover, regions with temperature distribution higher than 50°C at 1000 m deep, and higher than 90°C at 2000 m deep can be visualized.

Finally, the online tool provides information on the areas with potential for GeoDH and the heat-flow density. From the map, we can note the following:

- GeoDH can be developed in all EU-28 countries.
- The potential for GeoDH development by 2020 is much higher than the forecasts of member states in their NREAPs (see below).
- Geothermal can be installed with existing DH systems during extension or renovation, replacing fossil fuels.
- New GeoDH systems can be built in many regions of Europe at competitive costs.
- The Paris and Munich basins are the two main regions today in terms of number of GeoDH systems in operation.
- The Pannonian basin is of particular interest when looking at potential development in is Central and Eastern Europe countries.
- In southern Europe, the option of district cooling should be considered.

- The enthalpy (temperature) is not the only selection criteria; other key factors are heat flow on the supply side, and the heat users (urban density) on the demand side.

1.3.3.2. Geothermal heat pumps

Shallow geothermal energy is available everywhere, and it is harnessed typically by GSHP installations, using the heat pump to adjust the temperature of the heat extracted from the ground to the (higher) level needed in the building, or to adjust the temperature of heat coming from building cooling to the (lower) level required to inject it into the ground. The main technologies used to connect the underground heat to the building system comprise of:

- Open-loop systems, with direct use of groundwater through wells.
- Closed-loop systems, with heat exchangers of several types in the underground; horizontal loops, borehole heat exchangers (BHE), compact forms of ground heat exchangers, thermo-active structures (pipes in any kind of building elements in contact with the ground), etc.

For shallow geothermal energy (GSHP and UTES), the installation growth rate is even more spectacular, and a capacity of at least 17,000 MW_{th} was achieved by the end of 2012, distributed over more than 1.3 Mio GSHP installations.

The countries with the highest amount of geothermal heat pumps are Sweden, Germany, France, and Switzerland. These four countries alone account for ca. 64% of all installed capacity for shallow geothermal energy in Europe.

Looking at the time period 2010–2015, these four big players will have the greatest increase in terms of number of installations. In relative terms, Italy, Poland, and the Czech Republic are among the countries with the highest growth rate.

Today, the European-wide growth rate of the market for shallow geothermal systems was steady for some time, with new market actors filling the gaps left by others with decreasing sales.

The GSHP market today is in difficulty nearly anywhere.While in some mature markets the situation still is rather stable, in others a decrease can be seen. In parts of Germany, this can be attributed to continuously stricter regulation, causing delays and higher costs. Across Germany, and some other neighboring countries, GSHP systems are becoming less competitive; as the cost of electricity (which is required to run the heat pump) increases, the use of fossil fuels such as natural gas for heating becomes more favorable financially. In developing markets, the growth rate is low, minus 20% sales in some countries, and juvenile markets are not really progressing. Here the aftermath of the economic crisis and the low rate of construction in some countries take their toll.

With an economic recovery, a new increase in the GSHP market can be expected. Around 1 million geothermal heat pumps will be installed during the period 2010–2020. We can expect this figure to grow for 2020–2030 as more national markets become mature with growth rate higher than 10%. Legislations like the Energy Performance of Buildings Directive and the Energy Efficiency Directive will also have an impact on the future market of heating and cooling supply for new buildings and buildings renovation towards near zero energy buildings.

1.4. Perspective

This section will envisage the future development of the geothermal markets. The sector would grow by the accumulation of three factors:

- Removing barriers for a market uptake in all European countries and bringing to the sector the right support schemes.
- Developing the next generation of geothermal technologies and improving the current ones.
- Highlighting the role geothermal has to play in the energy union.

1.4.1. Barriers to Remove

Developing geothermal requires an enabling framework beginning with clear and consistent national and regional strategies from public authorities. From the project developer's point of view, realizing a geothermal project requires several authorizations and the compliance with a number of national and local regulations, and legal and financial safeguards. Regulatory barriers and long administrative procedures can result in additional costs. It is therefore crucial that a fair, transparent, and not too burdensome regulatory framework for geothermal is in place.

• Regulatory barriers

Regulatory and market conditions widely vary across Europe. In order to remove the regulatory barriers and promote the best practices, two main recommendations are set:

— Licensing procedures have to be simplified, and length and administrative burden have to be reduced.
— Market distortions and unfair competition from conventional sources (systems costs and externalities) must be compensated.

• Financial barriers

The geothermal energy source is free of cost, but the upfront investments to use it are significant. The higher upfront costs of geothermal can be compensated by much lower operating costs, but only if a sufficiently low "cost of capital" can be reached, that is to say if the risks can be properly managed. Therefore, innovative solutions for financing projects have to be found to overcome this challenge. Against this background, a combination of financing schemes and incentives can be a key point for the economic success of projects.

— A special focus has to be set on the geological risk insurance mechanisms that guarantee the presence and the quality of the resource. This could be a key aspect to overcome existing difficulties.
— When this important parameter has been overcome, in some cases, there is still a need for a comprehensive enabling framework in order to make geothermal competitive against fossil fuels

(as long as the final price of the latter does not fully reflect the real costs to society).

— Support measures for geothermal technologies are therefore needed to favor the progress toward cost competitiveness of a key source in the future European energy mix and to compensate for current market failures.

• Awareness

One of the main barriers is still the lack of awareness about this technology, its potential, and its advantages. Based on Europe's geothermal potential, geothermal energy could contribute much more significantly to the decarbonization of the energy sector. A considerable expansion of the geothermal sector is expected in the EU28 until 2050. For example, around 25% of the EU population are located in regions with hot sedimentary aquifers or other types of potential reservoirs, so are suitable for geothermal heating and cooling exploitation.

— Conditions to increase awareness about the advantages of this technology and its potential have to be created in each European country.

• Technical barriers

Although geothermal energy has provided commercial base-load electricity and heat around the world for more than a century, it is often ignored in national and European projections of energy supply. This could be a result of the widespread misperception that the total geothermal resource only relates to high-grade, hydrothermal systems that are too few and too limited in their distribution in Europe to make a long-term, major impact at a European/national level.

This perception has led to the undervaluing of the long-term potential of geothermal energy as the opportunity to develop technologies for sustainable heat extraction from large volumes of accessible hot rock anywhere in Europe has been missed.

Moreover, the potential of the geothermal industry can be achieved only through the attraction, retention, and renewal of the workforce. Companies and organizations need to team up to universities and research centers to shape and have access to the highly skilled workforce they need. Employment in the geothermal power industry is expected to increase while skill gaps and labor shortages may occur. For this reason, relevant public policy measures need to be consistent with energy policy (triggering change in employment needs) and industrial policy, and complemented by corresponding social and educational policies.

• Social and environmental barriers

Geothermal is fully recognized to be a safe, reliable, and environmentally benign renewable energy source. However, all human activities have somehow an impact on nature, including the construction of a geothermal plant.

Social acceptance is an important factor in site selection due to environmental issues (surface disturbances, noise, visual impact, reinjection, induced seismicity, stimulation and fracking), missing involvement issues, financial issues (in case of, for example, municipal grants), not in my back yard (NIMBY) acceptance issues, and local energy production.

The environmental impact of all infrastructure projects should be rightly considered, and environmental regulations are important tools for the development of geothermal.

Such a sustainable development of the geothermal sector would facilitate public acceptance. Lack of social acceptance can seriously damage the progression of geothermal developments and is an important issue to consider. Best practice shows that public acceptance is higher when project developers act openly and provide clear information, which helps to create trust.

1.4.2. New Geothermal Technologies

Research, development, and innovation will be much needed to develop geothermal technologies, accompanied by market uptake measures. The objective is to develop competitive sources of energy for the consumers and the industry. To this end, additional effort is need, particularly in the following areas:

• Market uptake of small-scale geothermal heating and cooling installations: Geothermal systems are already competitive in some markets in Europe. There is the need to remove barriers for a market uptake all over Europe.

- Innovation for allowing the fuel switch in DH and for industrial process: The industry and the DH sectors must switch to renewables such as geothermal; some innovations are needed for this transition: low-temperature systems, energy-efficient devices, etc.

- Demonstration of flexible geothermal power plants: For increasing the flexibility of the power system, flexible generation is essential and must be developed with renewable sources. Some renewables, including geothermal plants, usually run as base load, but new technology such as binary turbines allow them to be flexible in their production. More demonstration plants must be installed in different market contexts.

- Research and development of the next generation of RES technologies such as EGS: Breakthrough renewable technologies could be the future game changer for decarbonizing the energy system. EGS is a technology already demonstrated but an action plan must be launched for increasing its contribution to the electricity mix.

- Towards a smart integrated energy system: The future energy system should make a strong link between its three sectors: electricity, heating and cooling, and transport. Smart energy grids will play an important role in the future smart cities and communities by ensuring a reliable and affordable energy supply to various customers with renewable energy carriers like geothermal energy.

1.4.3. Its Future Role in the Energy Mix

We are surrounded by inexhaustible energy resources that allow us to meet our energy needs and those of future generations without taking uncontrollable risks with the life and well-being of our planet. The development of modern technology now enables us to make use of these energy sources on a scale that meets the requirements and demands of modern civilization.

A single technology, a single renewable energy can never meet this demand alone. Each alternative has its specific advantages and disadvantages, and has to be applied intelligently and targeted in those places where it can deliver its optimum strength in synergy with other technologies. Used in combination, renewable energy sources have a chance to meet the demand, especially for heating and cooling, which represents 50% of final energy demand in Europe.

Only a tiny portion of the potential geothermal energy is as yet explored and in use in Europe. Increasing the use of geothermal energy, and strengthening the geothermal industrial sector, will allow a substantial reduction of CO_2 emissions, the saving of primary energy, and the creation and sustainment of a work force with many skill levels.

The potential of shallow geothermal is significant. Often forgotten today, one of the main arguments to promote renewable energy sources in Europe is its local aspect. The local production of energy leads toward a decentralized approach and so a reduction of the system costs. It ensures also a security of supply with carbon free sources of energy. Having a local production of energy empowers the consumers who also become prosumers. The choice of the energy mix can then be more democratic.

Geothermal is a real local source of energy-producing power and heat for the cities and the rural communities. Geothermal allows also economical local development with many indirect positive effects such as jobs creation.

Moreover, one cannot look at an energy technology without considering its integration on the energy system and its role in the economy. And, here we can highlight the role of geothermal in the energy sector, and in the economy.

In the power sector, it will be a stabilizer of the electrical power on the grid providing firmness for the customers and ensuring security for the grid and the society. Geothermal power plants would be developed in all European regions. It should be taken into account in the current reform of the market design, adding a regional dimension between the centralized and decentralized approaches.

In the heating and cooling sector, geothermal will be one of the four renewables to compose the solutions of providing clean, competitive, and secure heating, cooling and domestic hot water for buildings and the industry. Geothermal has a key role to play in the decarbonization of this sector. The structure of the heating and cooling sector is more complex than the one for electricity. Many options would be available and the market will decide in each region the mix of energy sources. But both geothermal DH, direct uses and GSHP will become major technologies.

For ensuring this development, decision makers should go beyond the LCoE concept when preparing the energy mix. Today, this concept has gained too big an importance for both politicians and financiers although it does not allow the prices to reflect system costs and externalities as a burden paid by the society.

If the energy transition has to be successful, we have to think about an optimal scenario in terms of costs and affordability for the customers and the citizens. For geothermal, a local and stable source of renewable energy, the systems costs and the external costs are very much reduced. Moreover, it participates in the development of the local economy. For that, its role is crucial in the future energy system.

It is time now to go beyond the costs and the prices for the energy technologies and to consider their value. Geothermal benefits should be rewarded not only for its power and heat generation but also for its contribution to the energy transition.

References

EGEC, *Strategic Research Priorities for Geothermal Technologies* (2012).

EGEC, *Geoelec Final Report* (2013). www.geoelec.eu.

EGEC, *Geodh Final Report* (2014a). www.geodh.eu.

EGEC, *Geothermal Technology Roadmap* (2014b).

EGEC, *Regeocities Final Report* (2015) www.regeocities.eu.

EGEC, *European Market Report* 2015 (2016).

Chapter 2

Geothermal Binary Plants for European Development

Joseph Bonafin

Turboden, via Cernaia 10
Brescia 25124, Italy
joseph.bonafin@turboden.it

Geothermal projects are like fingerprints; each project is specific in its own way. This implies that some technical and economic prerequisites are required to understand the applicability of a power plant in a specific market.

Binary technology makes the majority of geothermal sources exploitable, due to low-temperature applicability and flexibility.

Organic Rankine Cycle (ORC) is the most widespread cycle for binary technology, with about 2 GW capacity installed and more than 500 projects in 80 countries currently under development.

The ORC principle is very simple: it converts the geothermal heat into electricity into a closed process, with no harmful emissions.

Geothermal power industry requires a multidisciplinary approach; in fact, many teams are involved in each phase of development: from geologist, to drilling experts, to geochemist, to the thermal and mechanical designer, to the financing partner.

Once the project has reached the feasibility phase and at least one production well proved the resource, the project's risk decreases and project financing is possible.

The operation of a binary geothermal plant is eased by high reliability and low maintenance costs.

Although a basic understanding of thermodynamics and heat transfer is required, the reader will be guided with simplified and practical approach to the binary power technology.

2.1. The Organic Rankine Cycle (ORC) Cycle for Geothermal Applications

2.1.1. Thermodynamics: Principles

In binary plants, there is no direct use of geothermal fluid for electric power generation, but a secondary fluid (the binary one) is heated up and vaporized by means of heat exchange with the geothermal fluid, in order to spin a turbine and produce electricity.

In geothermal power application, the general scheme is as follows (see Figure 2.1). A geothermal heat source delivers heat to a power plant that converts the heat (thermal energy/power) into electric energy/power. The heat that is not converted can be delivered to a thermal user, or dissipated by means of a suitable cooling system.

The net result of the conversion of the binary cycle (thermal power into electric power) is the net power, which is the difference between the gross generated electric power and the internal consumption of the auxiliaries like air-cooled condenser and pump. Minor auxiliaries like lubrication units can be neglected in a first evaluation. The ratio between electric power and thermal power is the electric efficiency.

$$\eta = \text{electrical power/thermal power}$$

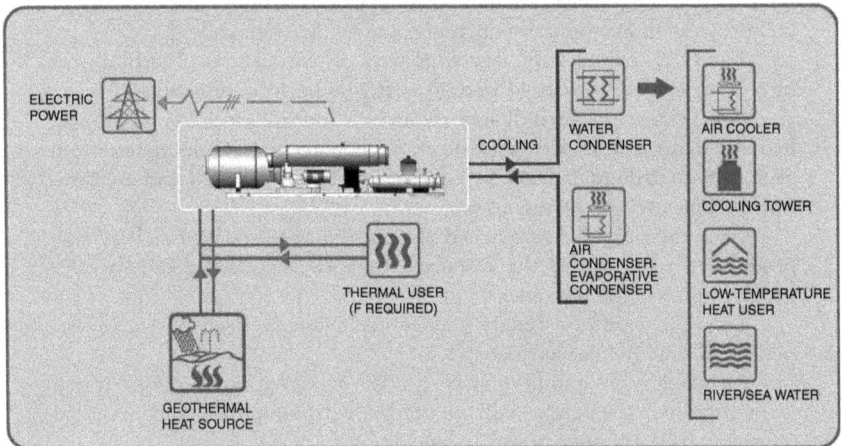

Figure 2.1   Geothermal binary plant scheme.

As a consequence of the previous formula, with a given geothermal source, the target is the electric power, to be maximized by means of a cycle designed is such way that the heat power introduction is the highest possible, and the efficiency of such heat introduction/conversion is compatibly the highest.

The heat is usually carried by a single-phase flow (either geothermal water or steam) or sometimes also by a two-phase flow (water and steam) by the pipelines of the gathering system, which collect the fluid from the production wells. The fluid, after cooling/condensation in the power plant, can be then carried back to the geothermal aquifer by the reinjection system. Typically, pumps are required either to produce or to reinject the fluids in the geothermal reservoir. Considering that thermal power has to be exploited in the long term (30 years), reinjection in the formations is preferable for environmental issues, and to preserve the level and pressure of the system.

As in the traditional Rankine cycle, the heat exchange drives the binary cycle (in many cases, an organic Rankine cycle), with conversion of the thermal power into mechanical power, thanks to appropriate items. First, the geothermal fluid preheats and vaporizes a suitable organic working fluid in the preheater and evaporator (2→3→4). The organic fluid vapor powers the turbine (4→5), which is coupled to the electric generator. The vapor is then condensed in the condenser, cooled by water or air (5→1). The organic fluid liquid is finally pumped (1→2) to preheater and evaporator, thus completing the sequence of operations in the closed-loop circuit (see Figure 2.2).

A quick way to calculate the electric power producible from a given geothermal heat source is to calculate the heat power available and estimate the efficiency of conversion. A general formula to calculate the thermal power (unit kW) available is indicated by the following formula:

$$\text{thermal power} = \dot{m} \times \Delta h$$

where $\dot{m}$ is the mass flow of the heat source (unit kg/s), and Δh is the enthalpy difference of the heat source between inlet and outlet of the power plant (unit kJ/kg).

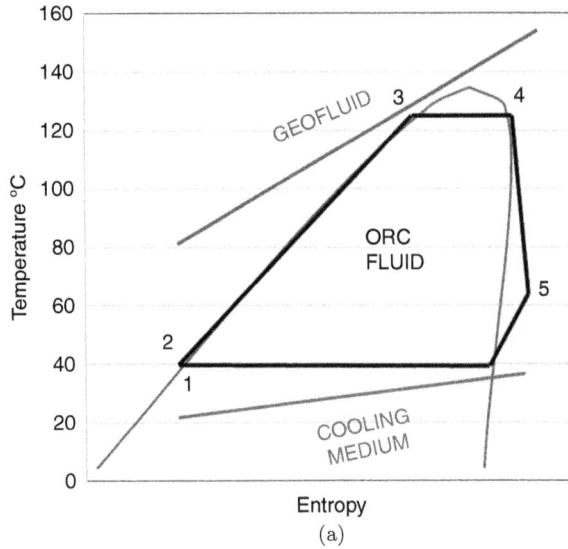

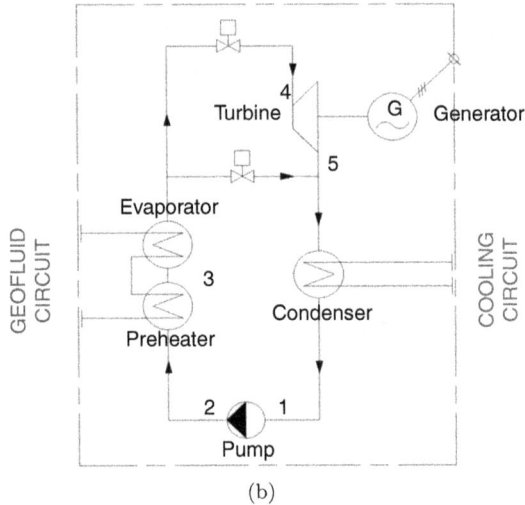

Figure 2.2   Temperature–entropy diagram of a binary cycle (a); basic process diagram (b).

For a binary cycle, a suitable way to set a benchmark for the efficiency is through comparison to the best ideal cycle.

Ideal means that the machine performing such a cycle has an infinite surface for heat exchange, and there is no irreversibility

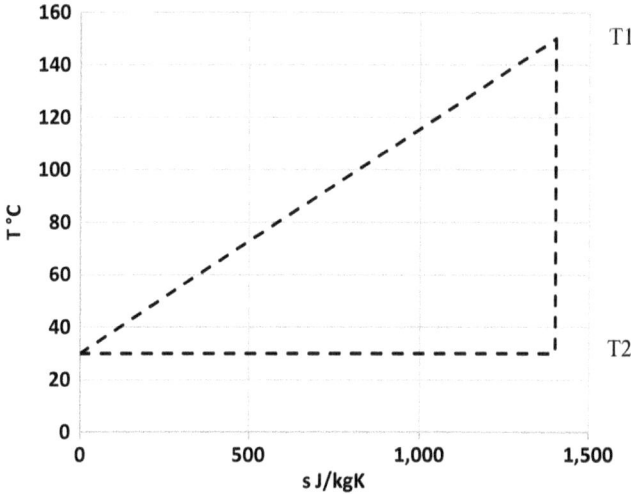

Figure 2.3   Temperature–entropy diagram of a "triangular" Lorentz cycle.

anywhere in the process (i.e., the turbine and the pump are assumed to perform isentropic transformations, with an efficiency of 100%).

Assuming no change of state (e.g., condensation/evaporation) in the hot/cold sources, an ideal cycle is similar to a triangular cycle consisting of an isobaric (constant pressure) heat addition process from the geo-fluid to the cycle's working fluid, up to the brine inlet temperature, followed by an isentropic expansion; an isothermal heat rejection process at heat sink's temperature completes the cycle. This cycle is known in thermodynamics as "triangular" Lorentz cycle, and can be represented by a triangle in the temperature–entropy diagram (Di Pippo, 2007).

In fact, the geothermal brine (in case of single-phase — water) is not an iso-thermal (or "latent heat") heat source, but it cools as it transfers heat to the working fluid.

The triangular Lorentz cycle (see Figure 2.3) efficiency formula is:

$$\eta_t = 1 - T2/((T1 + T2)/2)$$

where $T1$ is the geothermal water inlet temperature and $T2$ is the heat sink (ambient) temperature, both in degrees Kelvin.

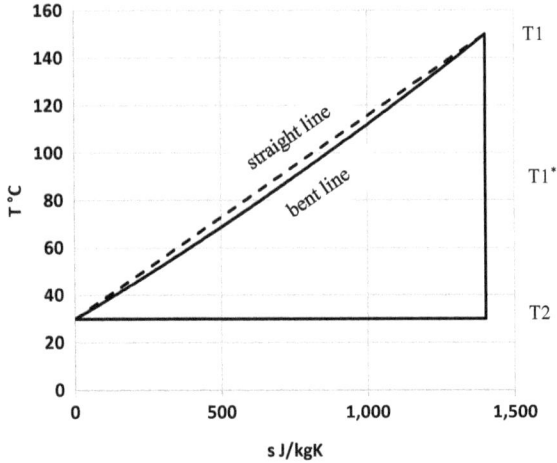

Figure 2.4 Temperature–entropy diagram of an ideal cycle with outlet temperature of the geo-fluid equal to the heat sink temperature, calculated with properties
of the water (bent line). The straight line is for comparison with the triangular
Lorentz cycle.

This is similar to the well-known Carnot efficiency formula (valid
for a rectangular cycle in the T–s diagram) but the denominator is
the arithmetic mean between $T1$ and $T2$.

There is an implicit approximation in assuming that the cooling
of a single-phase flow of water is represented by a straight line in the
temperature–entropy diagram.

In fact, by calculating the real properties in the T–s diagram of
the heat source during cooling, the shape of the ideal cycles becomes
like in Figure 2.4 and the efficiency can be more precisely calculated
by the formula of the "logarithmic triangular" Lorentz cycle:

$$\eta_s = 1 - T2/T1^*$$

where $T1^*$ is no more the arithmetic mean between $T1$ and $T2$ but
a suitable logarithmic mean between the two temperatures $T1$ and
$T2$ expressed by the formula:

$$T1^* = (T1 - T2)/\ln(T1/T2)$$

For example, if $T1 = 150°C$ (i.e., 423.15 K) and $T2 = 30°C$
(i.e., 303.15 K), the arithmetic mean (i.e., the mean value of the

temperature at which level the heat is provided in the process) is 90°C while the logarithmic mean $T1^*$ is 86.7°C, yielding to the following efficiencies:

- $\eta_t = 16.52\%$ (triangular Lorentz),
- $\eta_l = 15.75\%$ ("logarithmic triangular" Lorentz),
- $\eta_{ideal} = 15.84\%$ (calculated with the properties of the water).

Hence a difference of about 5% between the "triangular" and "logarithmic triangular" approach is quite significant, especially considering that the ideal efficiency (calculated with the properties of the water) is 15.84%, therefore the "logarithmic" efficiency method shall be preferred to set the benchmark of an ideal cycle efficiency, as it is a precise and conservative (underestimating) approach.

As it can be noted, the efficiency is mainly affected by the temperature difference between the heat source and the heat sink. Hence the seasonal variation of ambient conditions plays an important role in the overall efficiency.

If the hot source can not be cooled down to the heat sink temperature (as it is in most of the real power plants), the ideal cycle in the temperature–entropy diagram becomes a trapezoid (see Figure 2.5) with a concave upper line and the efficiency can be again simplified by the following expression:

$$\eta_{tz} = 1 - T2/T1^{**}$$

where $T1^{**}$ is the logarithmic mean between the two temperatures $T1$ and $T1out$ (respectively, the geo-fluid inlet and outlet temperature in K), and $T2$, expressed by the formula:

$$T1^{**} = (T1 - T1out)/\ln(T1/T1out)$$

For example, if $T1 = 150°C$ (i.e., 423.15 K), $T1out = 50°C$ (i.e., 323.15 K), $T2 = 30°C$ (i.e., 303.15 K).

$T1^{**} = 97.76°C$ (i.e., 370.91 K),
$\eta_{tz} = 18,27\%$ ("logarithmic trapezoid" Lorentz),
$\eta_{ideal} = 18,42\%$ (calculated with the properties of the water),
$\eta_{real} = 9.2\%$ to 11.0% (considering a real process with irreversibility).

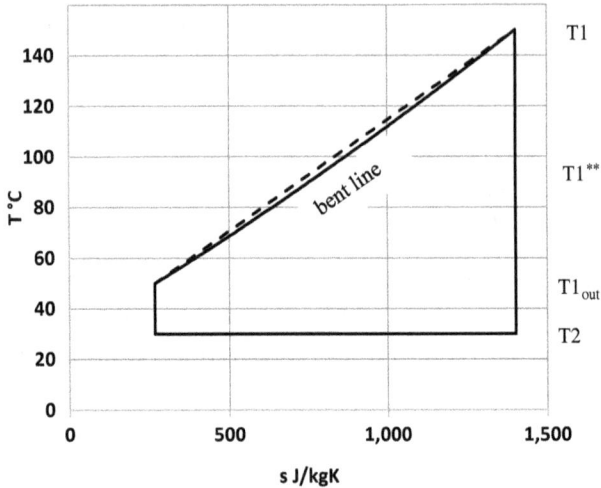

Figure 2.5   Temperature–entropy diagram of an ideal cycle with outlet temperature of the geo-fluid higher than the heat sink temperature, calculated with real properties of the water (bent line).

The increase in efficiency of η_{tz} compared to η_l is due to the higher mean temperature at which the heat is introduced in the process (i.e., $T1^{**} > T1^{*}$).

Real cycles with finite surface and irreversibility due to thermodynamic transformations in the equipment (e.g., heat exchange, compression, expansion) have in general (according to the current state of the art) a real cycle efficiency of about 50%–60% of the ideal cycle efficiency, i.e., for the previous example η_{real} is between 9.2% and 11.0%.

Therefore, given the geo-fluid inlet and outlet temperatures, and the ambient temperature, a quick tool to calculate a geothermal binary cycle efficiency, is to consider the 55% (i.e., the average between 50% and 60%) of the "logarithmic trapezoid" Lorentz efficiency.

Actually, also the heat rejection is normally performed to a variable temperature heat sink. The above-mentioned cycles assume rejection of the heat at ambient temperature, hence with an infinite surface of the air-cooled condensers and infinite flow of air. A further refinement of the efficiency calculation could consider calculation of the logarithmic mean temperature also for the heat sink, by considering to reject the heat at a temperature between 15°C and 30°C higher than the heat sink, to take into account the presence of a real air-cooled condenser of the binary plant.

As a consequence of these considerations, when a cycle has an efficiency of say 10% (meaning that for each 100 units of heat, 10 are transformed into electricity, i.e., gross power, and about 90 are dissipated to the heat sink), an improvement of only one percentage point represents a 10% improvement. This may make the difference if a project is economically feasible or not.

2.1.2. Limits of Feasibility

As a consequence of the Lorentz efficiency, the higher the temperature of the hot source (and the lower the rejection heat temperature) the more efficient/feasible is the thermodynamic cycle of a geothermal binary plant.

As a thumb rule, a project to be feasible must have at least 100°C available from the geothermal resource as inlet temperature of the plant. There is no technical upper-limit-temperature for a binary process, as the proper fluid selection can overcome any fluid stability issue (in principle, a binary cycle can use also water in its process).

Considering the lower-limit-temperature, even though it has been demonstrated that it is possible to exploit extremely low temperatures (i.e., 78°C [Low-Bin project in Simbach]), and below 90–100°C, the real cycle efficiency gets very low (<5% indicatively). Consequently, the heat exchanger surfaces needed to produce a certain electric power are proportionally larger due to the low efficiency of the cycle; also the heat sink's heat exchangers are larger due to the higher heat to be dissipated, hence the cost per kW of the heat exchange equipment increases, making the project hardly feasible.

On the other hand, having a high temperature (e.g., 200°C) available at well-head means having a powerful (high-enthalpy) source exploitable. In such conditions, binary power plants have the most traditional technology of flash plants as a competitor. In fact, as the enthalpy of the hot source increases (indicatively for values above about 1500 kJ/kg), the competitiveness of hybrid/flash systems increase.

The size of the power plant is another important discriminant in determining if a project is feasible or not. The market has demonstrated that small geothermal applications (few hundreds of kW capacity) are not fully sustainable. In fact, the risk of investment and cost for the exploration and drilling phases are too big for the economics of small plants. Typically, a project is sustainable at least above a 3 MW capacity (for an average value of the electricity of about 10 USDc/kWh). On the other hand, a "large size" project (i.e., hundreds of megawatts) is not a limit of competitiveness of binary plants against flash steam (or hybrid flash + binary) technology, provided the enthalpy is lower than about 1,000 kJ/kg.

2.1.3. Fluid Selection

Why selecting a fluid is different from water in binary cycles? First, it is a matter of temperature. In fact, by selecting a fluid with a lower boiling point it is possible to also exploit hot sources efficiently (like the geothermal one) within the range of 100°–200°C in order to heat up and evaporate with sufficient pressure a suitable working fluid, so it can be expanded through a turbine in order to produce work. Therefore, thermodynamic cycle configurations, which are inaccessible in the state diagram of water, can be obtained with organic fluids having different parameters, and with high efficiency.

There are other advantages such as selecting a suitable "organic" fluid instead of water:

- Dry expansion for many working fluids: No fluid condensation during the expansion occurs (unlike in the geothermal steam turbines), with no erosion of the last stages of the turbine, even with saturated cycle.
- Thanks to the higher molecular mass fluid, lower peripheral speed is required for the turbine, with lower mechanical stress.
- Lower enthalpy drop across the expansion in the turbine: Lower number of stages are required to efficiently convert the pressure into work in each stage, and turbine size and cost is smaller.

- Small expansion ratio across the turbine: It implies smaller blades size increase during expansion compared to water steam.
- Fluid pressure levels within the various components can be selected, to a certain extent, independently of the source–sink temperatures.
- It is possible to adopt optimum values for parameters that influence efficiency of the turbine such as isentropic head coefficient, specific speed, size parameter, and volumetric expansion ratio (Angelino *et al.*, 1984).
- Corrosive/erosive action toward metals (in heat exchanger and piping) is limited, due to the chemical characteristics and to the lower speed of sound.

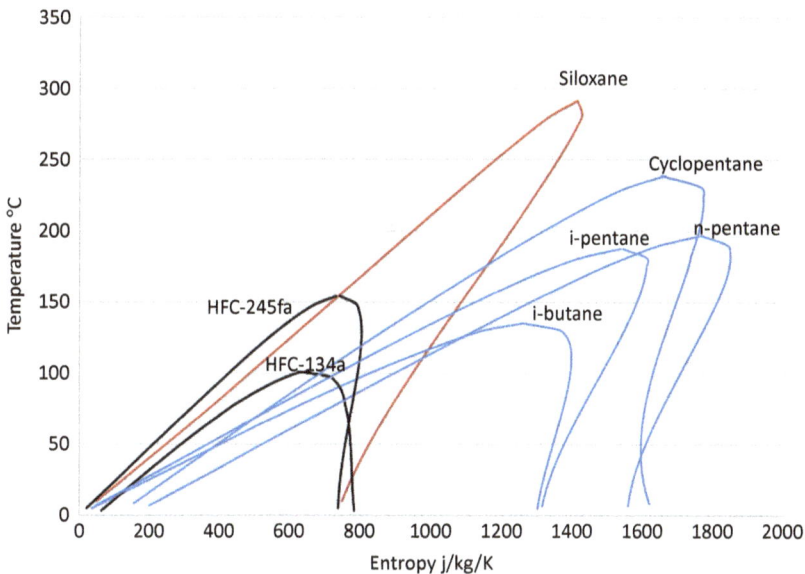

Figure 2.6   Saturation curves of the main organic fluids used in geothermal (a siloxane is plotted as a reference only), according to the temperature. The critical temperature is the point with highest temperature on each saturation curve.

Note: The ORC for higher temperatures like heat recovery and biomass use often different fluids like the siloxanes (see Figure 2.6), and a number of thermally stable aromatic hydrocarbons like toluene and diphenyl.

Figure 2.6 will help the designer to compare different working fluid cycles from the thermodynamic point of view. All the organic fluids plotted show a "dry expansion" steam region. As a thumb rule, the higher the critical temperature, the lower the ratio between internal consumption (for pumping the fluid) and gross power of the cycle.

In fact, in the comparison between two fluids (e.g., i-butane and n-pentane), it normally happens that one fluid provides higher internal consumption, but at the same time higher gross power. In such case, if there are no limits in the gross power generation, the solution that provides the highest net power must be preferred, notwithstanding the higher internal consumption to gross ratio.

The working fluid can be selected among a wide variety of alternatives available in the market, but only few of these are really usable for the power plant industry. Some limits of applicability of these fluids can be:

- toxicity,
- flammability,
- fluid stability (ability not to degrade during operation and cyclic thermal stress),
- global warming potential (GWP)/ozone-depleting potential (ODP).

In the light of the mentioned limits of applicability, the number of fluids are mostly restricted to the following groups:

- Hydrocarbons (propane, i-butane, n-butane, i-pentane, n-pentane, cyclopentane).
- Refrigerants (R-134a, R-245fa, and their substitutes at lower GWP such as R-1234).

How is a specific organic fluid selected in an ORC cycle? The aim is to convert the maximum quantity possible of heat into electric energy, by means of an optimum cycle, i.e., the net energy is the main parameter that must be maximized.

Figure 2.6 shows that it is possible to select among a number of different working fluids to design a cycle. However, the cycle that will convert more thermal energy into mechanical energy is, very likely, the cycle with the working fluid of which the critical temperature is closer to the temperature of the hot source. This is valid assuming a single-evaporation pressure level cycle. For number of pressure levels >1, a proper evaluation of the working fluid has to be done among the fluids with critical temperature similar or higher than hot source inlet temperature. In this regard, the outlet temperature of the hot source is the discriminant parameter.

Due to the Lorentz efficiency definition, the optimum fluid is the one featuring the maximum cycle-area in the temperature–entropy diagram, and minimum temperature difference between the heat source and heat sink temperatures.

In the example of Figure 2.7 (T–Q diagram), the geothermal heat source consists of two phases, with both steam and water available. The aim of the designer is to "inscribe" a thermodynamic cycle within the hot source (red ones) and cold source (blue one), respectively given by the geothermal specific source and ambient/cooling conditions. The area between the sources and the cycle is proportional to the exergetic losses of heat exchange (losses of power conversion capability) and must be minimized. A way to minimize this area is to increase the area of the heat exchangers up to the admissible pinch point limits. Another way is to select a different working fluid. A third way is to change the cycle configuration (e.g., more pressure levels). The T–Q diagram is therefore very helpful for the designer to have a quick qualitative comparison between a number of different potential solutions. In fact, not only it gives indication of the efficiency of the cycle, but also qualitative clue of the cost of the heat exchangers considering the pinch points and temperature differences. The principles of optimization will be explained more in detail in the following paragraphs.

Aiming to achieve the main goal through appropriate fluid selection, the designer must accept a number of secondary characteristics connected with the wanted target. The change of working fluid affects the design and performance of most cycle components, so that any judgment about the merits of a particular solution must include a quantitative evaluation of many different technical and economic

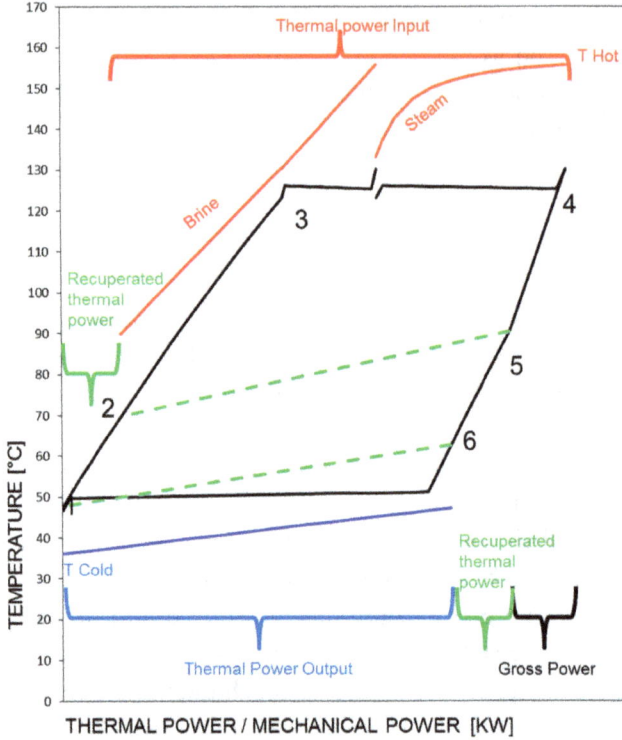

Figure 2.7   *T–Q* diagram for a binary cycle. Thermal power input is introduced in the cycle with cooling geothermal brine and condensing steam. A cycle is performed with recuperation of some heat at the turbine exit (5–6), preheating part of the fluid (1–2); this is beneficial to limit the cooling of the brine before reinjection, and reduce the thermal power output with advantage of the condenser. Gross power (4–5) is the difference between thermal input and thermal output. The ratio between gross power and thermal input is the cycle gross efficiency.

details (such as turbine efficiency, type of heat exchanger's geometry, design pressure, type of control of the cycle, cost of the equipment).

Finally, the designer must select the power plant that maximizes the net energy production according to the particular boundary conditions of the site. This shows that the design is not "static" but has to be considered the optimum solution among a dynamic environment. The most important variable to be considered for such energy optimization is the ambient temperature variation (seasonal), assuming that the geothermal conditions (temperatures and flows) are constant all over the operation time.

For a well-designed plant, with no significant off-design derating, the net energy producible in one year is circa the product of the design net power (at yearly average temperature), per numbers of operation.

2.1.4. Cycle Configurations

There are many possible cycle configurations for the same boundary conditions. However nearly all the configurations implemented so far in the binary power industry are either single pressure-level or double pressure-level cycles. Only few are made by alternative solutions (e.g., three pressure level cycle, cascaded multifluid cycles [Torbole ORC plant], supercritical cycles [Livorno ORC plant], Kalina cycles), and due to the relatively uncommon use, these solutions will not be described in detail.

The selection of the fluid is made already considering which configuration of the plant is the best one. In other words, the two evaluations (fluid selection and cycle configuration) must be done at the same time.

Considering the wide spectrum of combinations, the designer makes usually these comparisons to reduce the cases to be investigated (with increasing hot source temperature):

- Single-level propane VS single level 134a (from 100°C to 110°C).
- Single-level 134a VS double level 245fa VS double level *i*-butane (from 110°C to 140°C).
- Single level *i*-butane VS double level n-butane (from 140°C to 160°C)*.
- Single-level *n*-butane VS double level *i*-pentane (or *n*-pentane) (from 160°C to 200°C)*.

* *Note*: These qualitative comparisons can be applied to a liquid hot source, or in presence of a small steam fraction (typically <3%) that likely occurs for artesian resources for temperatures higher than 140°C. For resources with larger quantity of steam available, the comparison should include also the single level with pentane fluid.

Table 2.1 All the cycles are subcritical. (*) Comparisons have been made keeping the outlet temperature to the value that maximizes the net power output. This temperature is the one that makes the best compromise between efficiency of the cycle and heat input. As the source temperature gets higher, there is likelihood that the outlet temperature must be kept above 100°C. (**) Auxiliaries share considers the ratio between the auxiliaries consumption and gross power output. In the comparison, auxiliaries consumption considers an ORC pump with 75% efficiency, and ACC consumption equal to 1% of the dissipated thermal power.

Source temperature at ORC inlet	[°C]	110	110	140	140	140	160	160	180	180	200	200
Source temperature at ORC outlet (optimal)(*)	[°C]	61	60	42	58	55	70	63	74	74	54	76
Cycle configuration (pressure levels)		1	1	1	2	2	1	2	1	2	1	2
Regenerative configuration (yes or no)		No	No	No	No	No	No	Yes	Yes	Yes	Yes	Yes
Working fluid		Propane	HFC-134a	HFC-134a	HFC-245fa	I-butane	I-butane	N-butane	N-butane	I-pentane	N-butane	I-pentane
Gross electric efficiency		10,10%	9,10%	11,40%	11,20%	11,70%	15,60%	14,10%	17,40%	15,80%	17,20%	18,00%
Net electric efficiency		7,50%	7,10%	8,60%	9,80%	9,80%	13,00%	12,20%	15,00%	14,30%	14,60%	16,30%
Auxiliaries share (**)	[kW/ kW]	25,60%	22,30%	24,40%	12,50%	16,70%	16,80%	13,40%	14,10%	9,50%	15,50%	9,60%
Net power output per unit of mass flow rate	[kW/ (kg/s)]	15,6	14,9	35,7	34	34,9	49,5	50,4	67,3	64,2	89,5	85,2

The table illustrated above (see Table 2.1) shows typical values of gross and net efficiency, and specific net power per mass flow of hot source, considering an ambient temperature of about 15°C. These are the main indicators of how good the selected fluid-cycle-configuration is according to the boundary conditions.

As shown, the boundary conditions of a hot source are given, so that a suitable working fluid for the binary cycle allows to achieve a very good cycle efficiency with the single-level configuration, even better than the two-level configuration with a less adapted working fluid.

There are also other cases (not shown in the figure) where the percentage of steam is high enough to convert the latent heat of the steam quantity entirely in the evaporation heat for the working fluid. Also in such case, the single-level cycle can fit better than a two-level cycle the hot source, because of its "trapezoid" shape.

2.1.5. Optimization Principles

Once the best fluid and cycle configuration has been selected, the third step is to adjust the cycle parameters (e.g., working fluid flow rate and evaporation pressure) — by means of a sensitivity analysis — in order to finally select the design point and geometry of equipment. The main parameter in the optimization of an ORC power plant is the heat exchanger's surface.

Typically, a hot source is a single-phase liquid throughout the whole process of heat release. With reference to Figure 2.8, which represents the temperature–thermal power diagram of the heat exchange between the hot source and the binary fluid, if a subcritical cycle is adopted for the ORC fluid, the diagram shows a singularity called "pinch point," in which there is a minimum temperature difference between point 2 and 5 ($T2$–$T5$). The presence of a pinch point can be always determined in a heat exchange between two different media.

When the pinch point becomes very small (say below 5 K), the increase of heat exchange area becomes increasingly ineffective in terms of additional power. For example, for a single pressure-level cycle with 245fa as working fluid and a pinch point of 2°C, the increase of the heat exchange area up to 250% of the original value

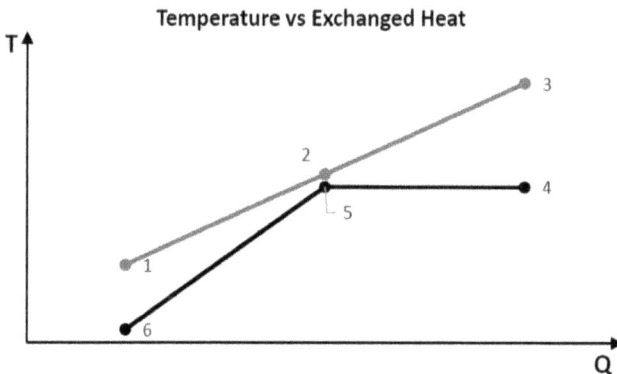

Figure 2.8   Temperature–thermal power diagram of saturated single-level cycle showing the pinch point (3–1 geothermal water cooling; 6–4 organic fluid heating and vaporizing).

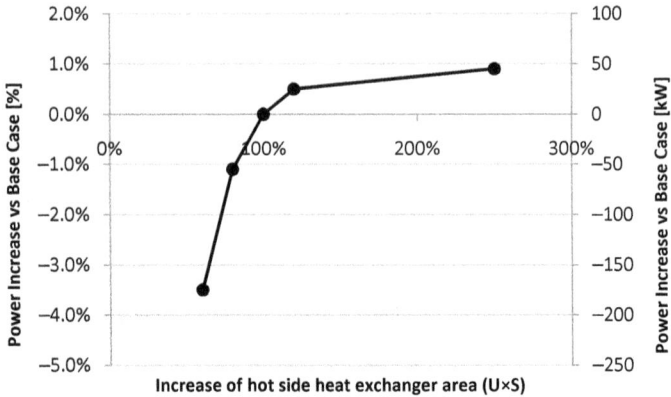

Figure 2.9　Influence of the increase (>100%) vs decrease (<100%) of heat exchange area on performance starting form a pinch point of 2°C.

would give a power increase of less than 1% (see Figure 2.9). It can, in fact, be demonstrated that a pinch point equal to zero leads to infinite exchange surface required. In other words, the cost of the heat exchangers to increase the power output over a certain limit is not justified.

In order to overcome this problem and reduce the exergy loss associated, a number of solutions have been considered in the past, like the adoption of proper zeotropic mixtures as working fluids in an ORC, or in a Kalina cycle, utilizing a mixture of variable concentration. Other ways are possible, like the shift to supercritical ORC as well as the adoption of multiple evaporation levels, or smoothing the pinch point (Gaia, 2013).

The designer must perform sensitivity analysis to evaluate which is the optimum cost of surface/net power ratio, according to the economic boundary conditions, to calculate the payback time and merit in the overall internal rate of return (IRR) of a more performing but expensive solution.

This indicates an important concept: neither the cheapest plant, nor the most expensive one in terms of cost/kW, is the most remunerative solution for a business plan. The optimum has to be discovered according to the specific case.

2.1.6. Equipment of a Binary Plant

This chapter briefly describes the general features of the main components of a geothermal binary power plant.

1. *Turbine*: At the heart of the plant, the turbine (one or two per single generator) is the component that transforms the energy within the working fluid into mechanical energy, spinning the electric generator. The optimization of this component is the key for maximization of the power output all over the wide range of seasonal and/or partial load operation.

The ORC turbines can be made with different geometry: axial flow (about 80% of the total capacity installed as of today by means of ORC turbines in operation in the world), centripetal radial flow (about 15%), or centrifugal radial flow (<5%) [*Source*: Turboden market survey]. Also, screw expanders have been experimented but with low availability.

The advantage of axial configuration is the possibility to optimize the isentropic efficiency over a wide range of operation, having the possibility to select within a wide range of working fluids and with high number of stages (at least up to 5) on a single shaft with the typical cantilever arrangement. With this state of the art equipment, sizes up to 20 MW per each shaft with isentropic efficiency higher than 90% can be reached. The revolutions per minute of axial turbines is typically the same of the synchronization speed of the generators (1,500 or 3,000 for 50 Hz, 1,800 or 3,600 for 60 Hz), thus no reduction gear is generally required.

The possibility to add a variable nozzle as a first statoric stage can be considered as an advantage for the operation of these turbines with variable upstream conditions (e.g., in presence of a district heating (DH) in parallel to the ORC power plant). However, it is possible to demonstrate that such variable nozzle does not give significant advantages during seasonal temperature variation (change in the downstream conditions).

2. *Condenser:* Part of the thermal power coming from the geothermal fluid is converted into mechanical power by the turbine, while the remaining part is dissipated by means of a condenser.

The condenser can be of a different type (see Figure 2.1), water-cooled or air-cooled (ACC). A water-cooled condenser needs a closed loop of cooling water, itself cooled by another item (wet cooling tower, or air cooler) or an open water loop with fresh water continuously provided by a river or a sea. Since water is not always available at site, the air-cooled solution is the most common. Also hybrid condensers such as wet and dry condenser in parallel, or dry and spray condensers, can be considered to combine the advantages of the dry (possibility to increase power output for temperature below zero) and wet (wet bulb temperature is significantly lower than dry bulb in hot and dry climate) cooling.

Many parameters can influence the design and the type of the air condenser and auxiliaries:

- **Working fluid:** According to saturation pressure at ambient temperature can be either above or below (under vacuum) the ambient pressure.
- **Air flow:** Induced flow or forced flow; typically the induced is selected for ORC application to minimize recirculation phenomena.
- **Number of tube passes:** Two passes fluid side is the best solution, in order to avoid the typical undesired subcooling effect traditionally present in the standard one-pass solution.
- **Tube bundles:** These are normally made in carbon steel, with finned tubes (different fin geometry are possible); material of the fin is usually aluminum. The air condenser is composed by a large number of bundles interconnected each other in parallel.
- **Design of the fans:** In order to optimize air flow *Vs* surface needed considering internal consumption and noise emission.
- **Noise limits:** Typically, the plants for continental Europe are close to inhabited areas, therefore the noise emission is more critical than in other areas and has to be controlled under the limits for all the sensible harmonics.

- **Safety:** In order to avoid leakage of working fluid, all the joints are welded, and no flanges are present. A solution with cylindrical welded bonnet can be preferred to the traditional configuration with plugged bonnet.

3. *ORC feed pump(s)*: The pumps can be either multistage centrifugal horizontal type, or vertical type.

Consumption mostly depends on the working fluid selected (see Table 2.1). The pumps are selected in order to guarantee the highest availability possible with at least one back-up pump (i.e., 2 × 100% or 3 × 50% or 5 × 25%). In case of failure of one pump, the control system is programmed to automatically switch on the backup pump without shutdown of the ORC plant.

ORC pumps are commonly driven by electric motors with variable frequency drive. This allows the pumps to change their speed according to the change in heat load input. The pump automatically follows the input by changing its flow (for an increase of heat load, the pump will increase its speed, and vice versa) in order to keep a certain set-point in the plant (either a working fluid liquid level in the evaporator or in the condenser or a superheating of the working fluid at turbine inlet).

It is also possible to regulate the flow of the pumps by keeping the speed of the pump constant and laminating the flow by means of a valve downstream the pump. This solution is obviously not preferable for high-pressure fluids, considering the worse effect on net power due to the associated losses.

4. *Heat exchangers and pipelines*: Heat exchangers have the function to transfer the thermal power from the geothermal water and/or steam to the ORC working fluid. Shell and tube heat exchangers often according to Tubular Exchanger Manufacturers Association standard are used. Geothermal fluid is carried inside the tubes, while working fluid is on the shell side. To avoid corrosion issues and to increase durability, the material adopted for the parts in contact with geothermal brine or steam (i.e., tube sheet, distributor channel, partition plate, and heat exchanger tubes)

can be a suitable stainless steel or duplex steel, or various grades of titanium, or carbon steel with a suitable anticorrosion layer. On the working fluid side, traditional carbon steel is used for the heat exchangers shells and for most of the piping. The system is equipped with shut-off valves able to isolate any heat exchanger from the rest of the circuit (e.g., for maintenance operation). Shut-off valves are installed on geothermal water side and on the working fluid side. In this case, only a small volume of the plant shall be drained during maintenance operation and the time spent to restore the plant in the normal configuration will be shorter.

5. *Recuperator* (regenerative exchanger): In case of relatively high exhaust temperature at ORC turbine outlet, it can be sometimes convenient to use the super-heated vapor to preheat the liquid after the ORC pump in a dedicated heat exchanger. Mainly (see Figure 2.7 and Table 2.1), the recuperator is useful to keep high efficiency when the optimum cooling temperature of the hot source would be below the minimum allowable reinjection temperature. If the temperature difference available from outlet of the turbine and liquid temperature is small, the heat that can be recuperated does not justify the presence of an additional heat exchanger.

2.2. Examples of ORC Power-Plant Configurations

In Europe, there are a number of proven geothermal resources suitable for power generation. From the south to the north there are Menderes and Canakkale in Turkey, Campi Flegrei and Larderello in Italy, Greek islands, Pannonian basin in Hungary and Croatia, Molasse Basin and Rhine Basin in Germany. Iceland can be included in these statistics as well, even though high-temperature steam fields are dominant. According to Bertani (2015), as of today, geothermal power plant capacity in operation counts more than 2 GW in Europe, of which 20% is made by Binary. Worldwide, about 14 GW are in operation, of which 60% are flash, 25% dry steam, and 15% binary technology. This shows how relatively low-temperature reservoirs in Europe are efficiently exploited by means of binary technology.

2.2.1. Mixed Steam/Brine Resources

If the geothermal resource is a two-phase water-dominated system like in the Turkish Menderes area, the vapor phase (steam + noncondensable gases [NCG]) flashed within the wellbore is separated from liquid phase before the power plant, by means of a suitable pressure vessel called "separator."

The brine and steam are sent in two dedicated bundles of the evaporator (see Figure 2.10). The condensate is collected and sent

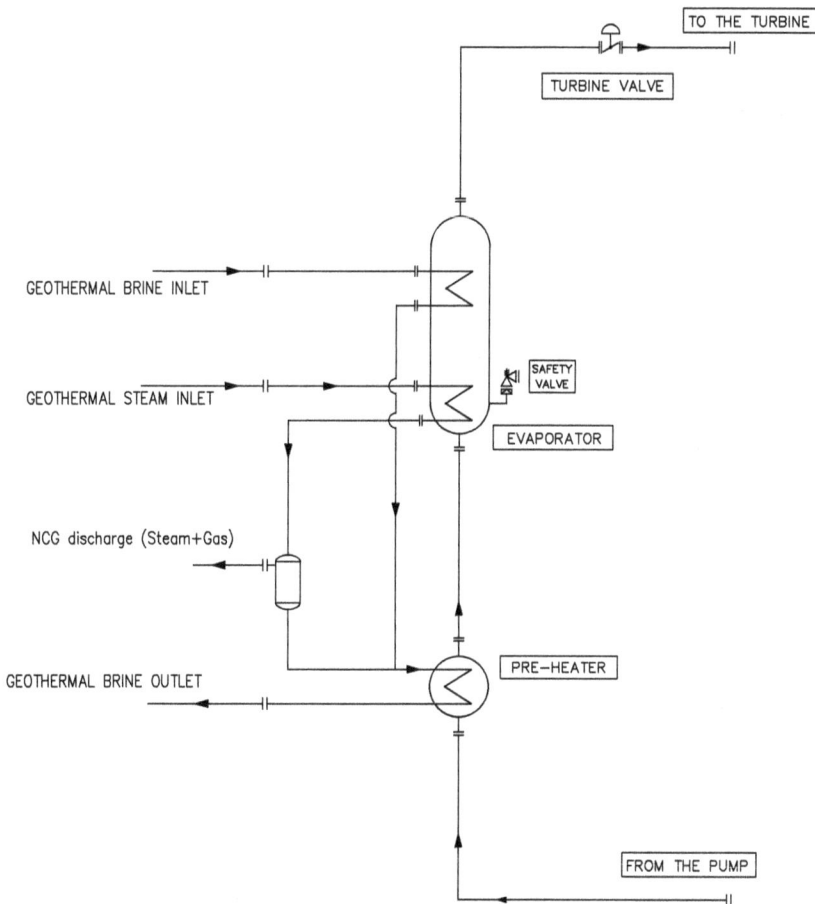

Figure 2.10   Binary plant process scheme using both steam and brine as heat carrier.

to the preheater mixing with brine at evaporator outlet (the mixing is not a requirement and shall be studied any possible chemical issue). An advantage of this cycle is that it has no sophisticated NCG removal equipment, as compared to the flash cycles. NCGs are separated from the condensate and can be vented (at pressure close to flash pressure) or expanded by means of an NCG expander in order to produce additional power.

Virtually, 100% of the produced geothermal fluid returns to the reservoir, other than some small water vapor loss carried with the vented NCGs. The high gas content can affect the heat transfer coefficients and profile of temperature within the heat exchangers and must be considered (see Figure 2.7 for the steam condensing at decreasing temperature, as the partial pressure of steam decreases while NCG partial pressure increases).

2.2.2. Combined Flash–Binary Systems

At locations where resource enthalpy and NCG content in geothermal fluid are high, efficient harnessing of the resource is a priority. Certain combined cycle configurations are able to overcome some of the limitations that accompany flash or binary cycles.

The use of such technologies may proliferate at sites with particular characteristics if they can be shown to be commercially competitive against more conventional approaches. The evaluation of appropriate cycles for a particular site should be conducted with geothermal experts familiar with the appropriate geochemical and plant process model tools that are capable of dealing with multicomponent mixtures, and the variety of technologies that are available to developers, in order to maximize return on investment and mitigate risk (Dunford *et al.*, 2014).

Combined Cycle (*Flash/Binary*)

This is the simplest configuration combining the two technologies. A separator produces high-pressure steam + NCG, which is used in a steam-condensing turbine that exhausts to a water-cooled condenser. Part of the steam is used to drive steam ejectors and keep vacuum condition in the condenser. Steam consumption is a function of NCG quantity. Alternatively, an NGC separator based on additional cooling (and condensing) of the residual steam has to be installed and the NCG shall then be compressed and purged away (or reinjected) with suitable compressors.

The brine from the separator is provided to a binary plant. The binary plant can be added also as a retrofit application to the existing single flash plant, if it is viable to further cool the brine before reinjection without scaling issues (see Figure 2.11).

Combined Cycle (F/B-B)

Figure 2.12 shows the proposed configuration of a combined cycle using two sets of binary units. Steam from the separator is sent to a backpressure steam turbine. This steam turbine exhausts at a pressure above atmospheric to a binary bottoming cycle ("Binary Cycle I") that works as a condenser for the steam cycle. This solution avoids investment in a NCG extraction system because of the high pressure kept in the condenser, and the overall net energy produced

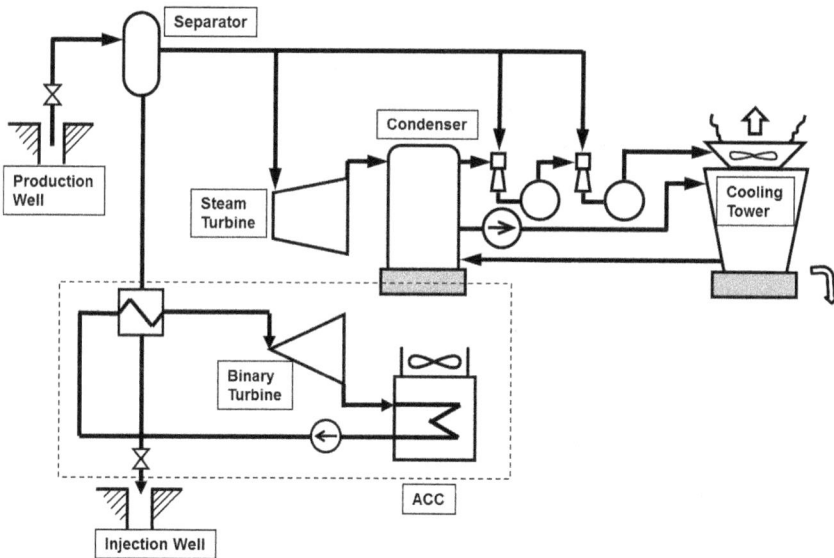

Figure 2.11   Combined cycle flash/binary.

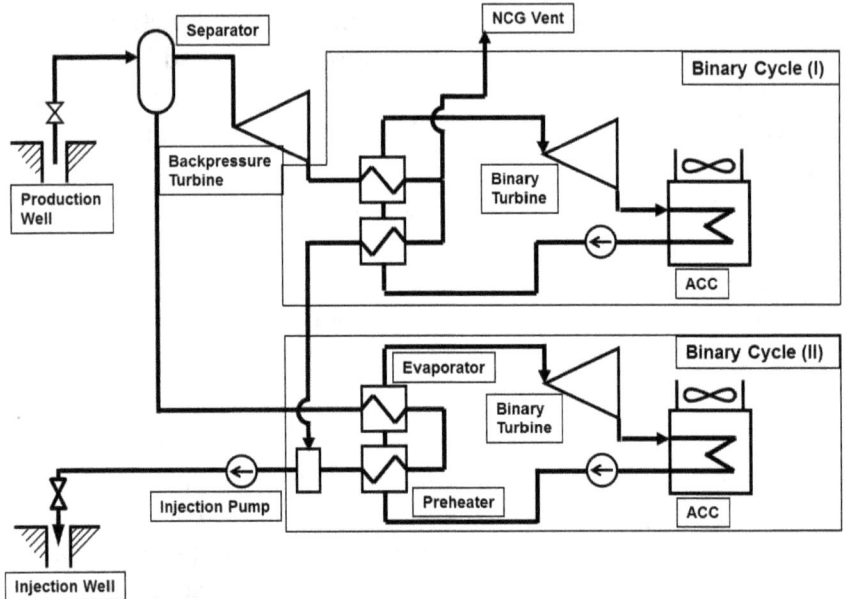

Figure 2.12   Combined cycle: flash/binary/binary (F/B-B).

by such F/B-B can be higher than a pure flash system because also the consumption of the NCG extraction system is avoided. Additional energy can be recovered from the hot water from the separator, and a separate binary cycle ("Binary Cycle II") is used to harness the energy. Such air-cooled combined cycle would be accompanied by similar advantages to the air-cooled binary unit, hence full reinjection of the geothermal fluid is possible. Similar systems are not yet present in Europe, but can be a good alternative especially for geothermal sources available in Italy and Turkey.

2.2.3. Nonartesian Resources

If the reservoir is deep (>3 km), the static level of water is likely below ground level. In such case, common to geothermal areas of Continental Europe (e.g., Molasse basin in Germany), a pump is needed to pressurize the water in the production well to feed the binary plant at temperatures above 100°C with a flow rate between 70 L and 150 L/s per well.

Only a few sites have sufficient artesian pressure at the wellhead to produce hot water without a pump. Even in this case, it can be necessary to use a pump for maintaining high pressure to avoid degassing and therefore precipitation of minerals (scaling).

Due to environmental reasons and to sustain the reservoir in the long term, the geothermal fluid has to be reinjected in the reservoir after the power plant; therefore, reinjection pumps at surface can be used to rise the pressure above the reservoir pressure.

Two type of pumps are used to produce water: line shaft pumps (LSP) and electric submersible pumps (ESP).

The LSP can be installed with a driver on the surface of the wells while a long shaft drives the pump impeller in the well.

The ESP are installed inside the well, with long electric cables to power the submersed motor.

As a simple comparison between the two different technologies: LSP generally have lower head/stage, less flow/unit/diameter higher temperature capability (up to 250°C) but shallower settings (down to 700 m). On the other hand, ESP have larger flow/unit/diameter, lower temperature capability (up to 150°C) but deeper settings (>1000 m). Also, LSP requires straight wells, while ESP can be installed in wells up to 75° off vertical.

These parameters should be taken in consideration in order to maximize the mean time between failures (MTBF): water level in the reservoir/drawdown, water temperature, discharge head pressure, gas quantity, well size (at least casing of 13 3/8″).

The pipelines of the balance of plant carry the geothermal water, after appropriate filtering, to the heat exchangers of the ORC plant. The water first passes inside the tubes of the evaporator, then inside the preheaters, transferring its thermal energy to the working fluid that is contained inside the pressurized shell. The water, cooled by the ORC heat exchangers to a temperature typically between 50°C and 70°C (depending on the specific chemical features of the geothermal fluid), is finally carried back to the aquifer by means of pipelines to the injection wells. The simplest configuration is made by a "doublet": one production well and one reinjection well. This system thus creates a closed circuit for heat exchange between the geothermal reservoir and the ORC plant, where the entire thermodynamic cycle of the working fluid takes place.

Until recently, geothermal power systems have only exploited resources where naturally occurring heat, water, and rock permeability are sufficient to allow energy extraction. An enhanced geothermal system (EGS) generates geothermal electricity without the need for natural convective hydrothermal resources. The permeability can be enhanced (engineered) by pumping high-pressure cold water down an injection well into the rock. Water moves through fractures in the rock, capturing the rock's heat until forced out of a second well as hot water. The heat is then converted into electricity using a binary power plant (Genter *et al.*, 2010). Although this system can be virtually applied everywhere the heat gradient is at least 3°C per 100 m (i.e., the standard value), the higher risk and cost of drilling has not still proven a mature applicability of EGS system.

2.2.4. Combined Heat and Power Geothermal Plants

According to European Geothermal Council market report (2015), there are 247 geothermal DH plants (including co-generation systems) in Europe. The total installed capacity amounts now to some 4500 MW thermal power.

According to the 204 planned projects, EGEC estimates that the capacity will grow to at least 6,500 MW_{th} in 2018.

The top five countries heated by geothermal energy are Iceland, Turkey, France, Germany, and Hungary.

When the temperature at wellhead is high enough to produce electricity, heat can be decoupled (i.e., the heat that is not used for electric production can be entirely or partly used for heat production) to a heat user (e.g., DH grid) with comparatively low impact on the electric power production.

From an environmental point of view, the saved carbon emissions are higher if the geothermal heat is used for heating purposes, due to the fact that the thermal efficiency for a geothermal heat source is up to 10 times higher when it is used for heating (100% thermal efficiency compared to 10–12% electric efficiency), while the same ratio is about 2 for fossil heat sources (roughly 80% thermal efficiency according to roughly 40% electric efficiency). Thus, geothermal heat substitutes much more fossil fuels if used for heating purposes than for power generation.

The value of the heat for electricity production in Germany is roughly €25/MWh if a feed-in tariff of €250/MWh and a net electric efficiency of 10% are considered. The economic value of heat, on the other hand, can be two times higher (€50/MWh), if used for DH (Bonafin *et al.*, 2015). However, a challenge for business plans based on DH only is the heat demand limited to the cold seasons.

In fact, while heat use is preferred from both environmental and economic perspective, suitable heat users are mostly available only for a limited number of hours per year (winter heating). Therefore, combined plants with efficient and flexible heat decoupling are very important.

Different schemes of heat decoupling can be implemented, mostly in parallel or in series with the power generation process. In the parallel scheme, the flow that is given to the DH is subtracted to the power plant (Traunreut plant), and a fully integrated control system is implemented with priority to the hat production. This scheme is effective when the heat peak load is high (>30% of the total heat available). On the other hand, heat can be cascaded to a user (i.e., connected after the binary plant and at lower temperature), when the electric power production is a priority or the DH not so large.

Compared to a single-level cycle, the two-level cycle is particularly suitable for the decoupling of heat with comparative lower impact on the electric power production. For example, in the double pressure-level of the Sauerlach binary plant, the heat supply can be connected downstream of the high-temperature loop, therefore decreasing mainly the heat supplied to the low-temperature loop (i.e., the cycle with lower electric efficiency). As a result, the energy loss when the heat is required for DH will be significantly lower.

Co-generation with heat decoupled at condenser of the binary plant is not a viable solution considering the inherent low temperature required by heat sink of the system.

2.3. Economics of Binary Geothermal Plants

The economic performance of a geothermal project's business plan is mainly influenced by:

- The cost per kW of the equipment. Generally, the higher the geothermal source temperature, and the bigger the size, the lower it is the specific cost of the equipment. For large-high temperature plants (e.g., 20 MW at 150°C) to small low-temperature plants (i.e., 3 MW at 110°C), the cost per kW of the binary equipment can, respectively, vary between €1000 and €2000 per kW installed. Besides the temperature of the source, a low average ambient temperature increases the power production hence significantly contributes to lower the cost per kW of the equipment and to increase the annual energy production. The use of stainless or duplex steel to be used in the heat exchangers due to an aggressive quality of the geothermal source also contributes to increase the price per kW of the unit, compared to a carbon steel solution.
- The cost of drilling: Typically with cost in the range of €1,000 to €2,000/m but also with an associated risk to overflow the budget for unexpected problems during drilling.
- The resource availability at a shallow depth means lower operation cost for wells and pumping systems, including reinjection costs.
- Well productivity: The required flow rate has to be achieved without causing too large drawdown, otherwise the parasitic load for running the production pump will be too high.
- The capacity factor, or availability (running time per year) of the system at the design capacity. For geothermal projects, this value is on average 90%.
- The value of the electricity sale per kWh produced; a power purchase agreement (PPA) with fixed value for at least 10 years of operation is attractive for investors.

- The opex value, which is generally in the range of €5–€15/MWh for binary plants, including all the operation costs, maintenance of wells and equipment, fees.
- The resource "aggressiveness" that can cause scaling or corrosion problems. These issues can require frequent cleaning of the system and/or intensive chemical dosing of the geothermal fluid.
- Macroeconomic aspects, like the weighted average cost of capital, or political risk of country.

As a rule of thumb, capex of a binary geothermal project is so divided: one-third for the exploration phase and drilling cost, one-third for the equipment of the power plant, and one-third for the balance of plant and construction activities.

2.3.1. Project Phases

Financing of geothermal projects is more difficult in the earlier stage of development. Factors such as geological potential, flow rate, market and political environment greatly differ from place to place. Financing also largely depends on local energy markets, and public and political support. This reflects the risks that geothermal projects bear until the resource is proven. The further the project develops, the more the risk decreases and the financing becomes easier. The lack of capital available during the early stages of development has hindered the growth in the development worldwide. Most geothermal projects take 5–7 years to be fully operating, depending on permitting and other licensing issues. Usually 3–4 years until feasibility phase, 1 year for drilling, and 2 years for above-ground equipment construction and installation. Financial applications greatly depend on the success of drilling and the resources available. The drilling success is the proven resource in terms of flow, temperature, and pressure of fluids. The above-ground technology is the one with lowest risk in the development chain, and also the one with shorter delivery time.

A geothermal project encompasses different phases. Each phase requires different equity and financing solutions with very different risk profiles at each stage.

- Site identification phase.
- Prefeasibility phase — surface exploration.

- Feasibility phase — exploration drilling.
- Resource development phase — production drilling.
- Plant construction phase (and completion of reinjection drilling).
- Operation and maintenance phase.

Generally, geothermal binary projects have a payback time between 5 and 10 years (from the start of the drilling operations, thus excluding all the prefeasibility and exploration phases whose time and cost are very site-specific).

Project's IRR percentage is generally higher than 15% @ 20 years, the typical reference time considered in business plans.

In markets with long horizon for investments (e.g., Turkey), geothermal industry has grown dramatically in the last five years, thanks also to specific financial support schemes that further increased the project's remunerability.

2.4. Operation of a Binary Plant

During the operation, all the relevant parameters of a power plant are monitored from remote by the SCADA system, to support the decision when periods of maintenance become necessary. Typically, a daily check is enough, and regular (yearly) maintenance is made to verify and overhaul (if necessary) vital components of the system. Binary systems have few rotating parts subject to mechanical wear (e.g., turbine and pumps' seals or bearings), and few vital signals that are redundant in order to keep in operation the control loops in case of instrument failure or malfunctioning. Electric parts require the typical maintenance of any power plant application. The binary circuit is technically tight and welded, so that only minor losses of working fluid occur, with a typical annual rate of loss <0.5% of the total quantity. An overhaul of the turbine can be conservatively considered after 10 years of operation. Cleaning of the heat exchangers and ACCs is a required periodical activity on the binary plant to reduce the fouling issue (besides some activities in other parts in contact with the geothermal fluid external to the binary plant itself), and can be only determined by controlling the decline of energy production over a period.

Drilling of make-up wells is included in the O&M costs, especially in larger high temperature/shallower resource plants. These wells are required either to reach the planned capacity (when the plant size has been already fixed before all the required production wells are realized) or to maintain the full generation capacity as there will be a slow decline in productivity/temperature of the wells with time.

In case of low-enthalpy doublet systems, the maintenance and possible replacement of the production pump is a major issue. For very deep reservoirs, developers should conservatively consider in the business plan a contingency to include a spare ESP pump for each production well, and a substitution of the pump once a year, with downtime of two weeks. Alternatively, redundancy can be implemented by adding a second production well and reduce the flow and pump load in each well.

Hot geothermal fluids contain concentration of NCG and dissolved solids. Scaling must be prevented from calcite and silica deposits from geothermal fluids. Also stibnite can be encountered as a type of scaling, especially in binary plants. Such phenomena must be avoided, otherwise part of the piping or wells can be completely compromised in short time. In fact, scaling deposits reduce the diameter of the wells and can even clog the reservoir. A proper design of inhibitor systems and pressures in the circuits is required to avoid calcite scaling, while temperature control during cooling of the brine is the best countermeasure not to have supersaturated silica.

Some geothermal fluids are acidic, and would rapidly cause corrosion to the parts in contact with the fluid (i.e., well's casing, piping, bundles of the heat exchangers). When it is not possible to add alkaline solutions, special steels (e.g., stainless steel or duplex steel) must be employed or suitable corrosion layers (or coatings) must be considered in the inner surface of the pipes.

2.4.1. Advantages

Operational costs strongly vary for different plant types regarding the size, the quality of geothermal fluid (scaling, corrosion, fouling etc.), and reservoir conditions. However, while the power plant is in operation, it creates base load sales revenue from the PPA and/or from the sale of the heat (in co-generative plants), or the sale of by-products like CO_2 for food industry.

It is relatively easier from an O&M perspective, as well as for monitoring and safety reasons, to deal with geothermal plants:

- Fluid pressures and temperatures are very low (compared to traditional generation systems).
- Geothermal plants' energy production is dispatchable and provides a base load.
- The plant can be remotely and unattended controlled.
- Process control is simple and fast.
- Maintenance procedures are quick (one week per year on average).

In case of failure of the grid, the ORC plant island mode can be adopted to supply the power required by all auxiliaries (included the geothermal pump), thus maintaining the plant in operation (Sauerlach plant, Turboden). The island mode presents a list of advantages, the significant of which are the following:

- The whole plant is maintained in operation and ready for a new reconnection with the grid.
- Less time is spent by the operators to check and restart the plant.
- Limitation of the number of start and stop cycles for the geothermal pump.
- Limitation of large precipitation of dirt along the well.
- Avoid the interruption of the flow in the pipeline with precipitations inside the tubes and filters.
- Increased availability of the delivery of the heat to the DH.

For these reasons, the availability is on average >90% and the costs of operation per energy produced among the lowest in the power industry.

Levelized cost of electricity (LCOE) is often cited as a convenient summary measure of the overall competitiveness of different generating technologies. It represents the cost per-kilowatthour of building and operating a generating plant over an assumed financial life and duty cycle. Key inputs to calculating LCOE include capital costs, fuel costs, fixed and variable operations and maintenance (O&M) costs, financing costs, and an assumed utilization rate for each plant type.

On the other hand, levelized avoided cost of electricity (LACE) provides an estimate of the cost of generation and capacity resources displaced by a marginal unit of new capacity of a particular type, thus providing an estimate of the value of building such new capacity. When the LACE of a particular technology exceeds its LCOE at a given time and place, that technology would generally be economically attractive to build. According to International Energy Agency (IEA) estimate of generation resources, geothermal is in many cases not only the resource with lower LCOE, but also the only with positive difference between LACE and LCOE.

2.5. Conclusions

A wide overview of binary ORC systems for geothermal application has been presented. Such systems are generally flexible for a variety of heat sources. In particular, binary ORC can be the best solution (or even the only alternative) to exploit low-temperature sources as in the case of Europe. Technical and optimization features have been explained, with direct reference to existing plants in operation. Investment in such projects can be long, but with good returns and low operation risk. Geothermal energy is base load, with no cost of fuel. Technological suppliers proved the availability of mature and reliable standardized systems. Despite the limit of low-temperature application, innovation and research can still introduce improvement margins. Support schemes are still required to boost development and mitigate the risks, typically during the initial phase. A global growth of this renewable energy must be expected in the next decades, with exponential rate.

References

Angelino, G., M. Gaia, and E. Macchi, A Review of Italian activity in the field of Organic Rankine Cycles. Proceedings of International VDI Seminar on

"ORC HP Technology Working Fluids Problems". Zurich. September 1984. VBD Berichte 539. pp. 465–482.

Bertani, R. Geothermal Power Generation in the World 2010–2014 Update Report. Proceedings World Geothermal Congress 2015. Melbourne, Australia, April 19–25, 2015.

Bombarda, P., M. Gaia, C. Invernizzi and C. Pietra. Comparison of Enhanced Organic Rankine Cycles for Geothermal Power Units. Proceedings World Geothermal Congress 2015 Melbourne, Australia, April 19–25, 2015.

Bonafin, J., M. Del Carria, M. Gaia and A. Duvia. Turboden Geothermal References in Bavaria: Technology, Drivers and Operation. Proceedings World Geothermal Congress 2015. Melbourne, Australia, April 19–25, 2015.

Di Pippo. Ideal thermal efficiency for geothermal binary plants, *Journal: Geothermics* Volume 36, Issue 3, June 2007, Pages 276–285 Initial of Author: R (2007).

Dunford, T., B. Lewis, K. Wallace and W. Harvey. Power Engineers. Combined cycle strategies for high enthalpy, high non-condensable gas resources. *GRC Transactions*, 38, 1 (2014).

EGEC market report. http://egec.info/wp-content/uploads/2011/03/EGEC-Market-Report-Update-ONLINE.pdf.

Foresti, A., D. Archetti, and R. Vescovo. ORCs in steel and metal making industries: lessons from operating experience and next steps (Ref. Torbole plant).

Genter, A., X. Goerke, JJ. Graff, N. Cuenot, G. Krall, M. Schindler and G. Ravier. Current Status of the EGS Soultz Geothermal Project (France). Proceedings World Geothermal Congress 2010. Bali, Indonesia, April 25–29, 2010.

Livorno plant. http://www.asme-orc2013.nl/uploads/File/PPT%20166.pdf.

Low-Bin project. http://www.lowbin.eu/index.php.

Traunreut plant. http://www.geothermie-traunreut.de/.

Chapter 3

The Icelandic Experience on Integrated Geothermal Utilization

Ólafur G. Flóvenz* and Brynja Jónsdóttir

ÍSOR, Iceland GeoSurvey, Grensásvegur 9
Reykjavík, Iceland
* ogf@isor.is

3.1. Introduction

Iceland is, as the name of the island indicates, a country of relatively cold climate located just south of the Arctic Circle in the North Atlantic. It has, however, a typical island climate since a part of the Gulf stream of the North Atlantic Ocean surrounds the country and keeps it relatively warm in the wintertime, compared to continental Northern Europe, but relatively cold in the summertime. The average temperature in January in the lowlands is just below zero but barely reaches 10°C in the hottest month of July. In wintertime, cold polar winds can bring the temperature down to −20°C for a few days in extreme cases. The low average temperature houses more or less all year around, albeit the average energy consumption in the coldest winter months is at least twice as high as in the summer.

Iceland was uninhabited until the 9th century when thousands of people emigrated from Norway and the Celtic nations, settled in Iceland, and established a state with a parliament but without a government. The settlement period lasted from around 870 to 930 A.D. and was possibly terminated by the consequences of a huge volcanic eruption in the year 934 (Hjartarson, 2015). The people mostly lived from animal farming supplemented by coastal fishing. During the initial settlement period, the country was covered with forests and bushes of birch and willow, which provided wood for cooking and heating. However, aggressive overgrazing and nonsustainable utilization of the forests caused most of them to disappear in a few centuries.

The nonsustainable land utilization, slight climate cooling, and a few large pyroclastic eruptions caused the life in Iceland to be difficult until in the 20th century. The population of the country had been averaging around 50,000 throughout the centuries but started to grow in the late 19th century and is now 330,000 in the year 2016.

From a geological point of view, Iceland is a unique country, being the only country on the planet where active ocean spreading ridge is above sea level. The geology of the country is dominated by ocean floor spreading, extensive volcanic and earthquake activity that creates abundant geothermal fields with hot springs and fumaroles. This chapter describes how geothermal energy has been harnessed and how it creates the geothermal culture in the country that is a part of the everyday life of the Icelandic people.

3.2. The Icelandic Geothermal Culture

Large-scale utilization of geothermal energy in Iceland has developed during the past 90 years or so. It has been growing steadily, after a slow beginning when technical problems had to be solved and nontechnical barriers overcome.

Before the 20th century, life in Iceland was a struggle, with the cold and harsh climate and lack of energy to heat the houses. There were hardly any villages and most people lived permanently on small sheep farms in the countryside but part of the year in poorly uninsulated shelters at the sea side during the fishing season. The country was almost devoid of forests and peat was the main source of heat. The houses in the countryside were mainly built of loose stones bound together with turf and covered with grass for insulation. The windows were very small to reduce heat loss. When urbanization started in the late 19th century, the houses were built of imported timber, but with poor insulation and heated mainly by imported expensive coal. The result was poor living conditions due to low indoor temperature during wintertime and outdoor pollution from the coal burning. Economically, Iceland was one of the poorest countries in Europe.

This changed dramatically in the latter half of the 20th century when Iceland developed very rapidly from a poor country to a wealthy one that can offer a comfortable life. There are several reasons for this change, but among them is the successful development of cheap geothermal energy for various uses, thus making life much easier, even during long, cold, and dark arctic winters.

Nowadays, geothermal energy is an integrated part of everyday life in Iceland and has created a particular culture that marks the society in many ways. About 90% of all buildings in Iceland are heated directly by geothermal energy, both in urban and rural areas and in most cases at very low price. The abundance of hot water affects everyday life. All parts of the houses are heated throughout and the typical room temperature is 22°C, even in the garage. You can take a long hot shower with high flow rate without worrying about your energy bill. And you can, without any additional energy cost, use the return water from your radiator system to heat up the pavement outside your house and keep it free of snow. You can go to the one of the numerous outdoor warm and comfortable geothermally heated public swimming pools and enjoy swimming, both on beautiful summer days as in winter snowstorms; or enjoy sitting outdoors in a hot tub with friends, participating in discussions and debates on politics or simply the life. You can also visit geothermal spas or buy fresh vegetables throughout the year, which are grown in geothermally heated greenhouses under lightning produced by renewable electricity. People who enjoy sports can utilize some of the numerous heated sports halls for exercising. In recent years, six full-size geothermally heated indoor football grounds have been built, and also numerous full size sports halls for handball and basketball, providing excellent facilities for the development of youth sport skills. The utilization of cheap geothermal energy for domestic heating is one of the reasons for the international success of the Icelandic national teams in handball and football.

It should be mentioned that geothermal energy provides more than purely economic value. It is intertwined with the Icelandic culture, at least with the Icelandic language. The first recorded permanent settler in Iceland, Ingólfur Arnarson, decided to build his home and settle down in a creek where he saw steam rising from the ground. He named the place Reykjavík, which is a compound word, *reykur*, means smoke or steam and vík, which means a creek. Many other places where hot springs are found also include the word *reykur*, like *Reykja-nes*, *Reykir* and *Reyk-hólar*. Other ancient names all over the country refer to geothermal energy. Geyser is the international word for an erupting hot spring and is derived from *Geysir*, the name of the greatest eruptive hot spring in Iceland. The Icelandic name *geysir* is derived from the Icelandic verb *gjósa*, which means to erupt. A large number of ancient place names refer to geothermal activity. The word *laug* means a warm spring that can be used for bathing and washing. It is very common in place names like *Lauga-land*, *Lauga-ból*, or *Laugar*.

The word *hver* means a natural hot spring of bubbling water, frequently close to boiling and is usually substantially hotter than *laug*. It is reflected in place names like *Hvera-gerði* and *Hvera-hlíð*. Finally, the street *Lauga-vegur* in Reykjavík, presently one of the main shopping streets in the city center, was originally about a 4 km long street from the city center to the hot springs in the valley *Laugar-dalur*, where women in Reykjavík used to go carrying clothes to wash in the hot springs.

3.3. Geothermal Resources of Iceland

3.3.1. The Geology of Iceland

The geology of Iceland is unique as the island is the only place in the world where an active oceanic spreading ridge is above sea level. The reason for this is the presence of a low-density mantle hot spot, centered below Iceland, which increases the magmatic production rate compared to normal oceanic ridges (e.g., Wolfe *et al.*, 1997). This leads to abnormal crustal thickness beneath Iceland of 20–40 km (Bjarnason *et al.*, 1993; Kaban *et al.*, 2002) compared to normal oceanic crust of 10 km or less.

The spreading axis of the Mid-Atlantic Ridge (MAR) crosses the island as a zone of active spreading and volcanism, referred to at the axial volcanic zone (AVZ). The measured half spreading rate in Iceland is close to 1 cm/year. The axis rises from sea on the Reykjanes peninsula on the south-west corner of the country and submerges again at the north-eastern coast (Figure 3.1). The AVZ does not form a straight line through the country but is shifted 150 km eastward, close to the southern coast and back in a westerly direction at the northern coast. In the southern part of Iceland, the AVZ is composed of two parallel axial segments. Throughout the almost 20 m.y. of the exposed geological history of Iceland, the AVZ has shifted eastwards a few times, leaving behind traces of ancient spreading axis and related transform tectonics, especially on the American plate (e.g., Sæmundsson, 1979).

The subaerial volcanism of the country resulted in extensive eruption of flood basalts that characterizes the preglacial and interglacial times while during the glaciation periods, when the country was mostly covered with ice, elongated hyaloclastite ridges or table mountains were formed above the volcanic fissures. The glaciation and deglaciation furthermore lead to large vertical crustal movements that might have contributed to the formation of fracture dominated hydrothermal fields outside the rift zones (Bödvarsson, 1982). The last glacial period in Iceland ended about 11,700 years ago (Cohen *et al.*, 2013).

The crustal accretion process in Iceland has been modeled and described by Pálmason (1973). His model assumes a simple spreading axis and fixed spreading rate, where new crust is partly formed by dyke injection into the existing crust and partly by surface volcanism. The eruptions cause the lava to accumulate at certain rate, normally distributed around the spreading axis, and the crust subsides by the same amount as the overlying erupted mass. This means that

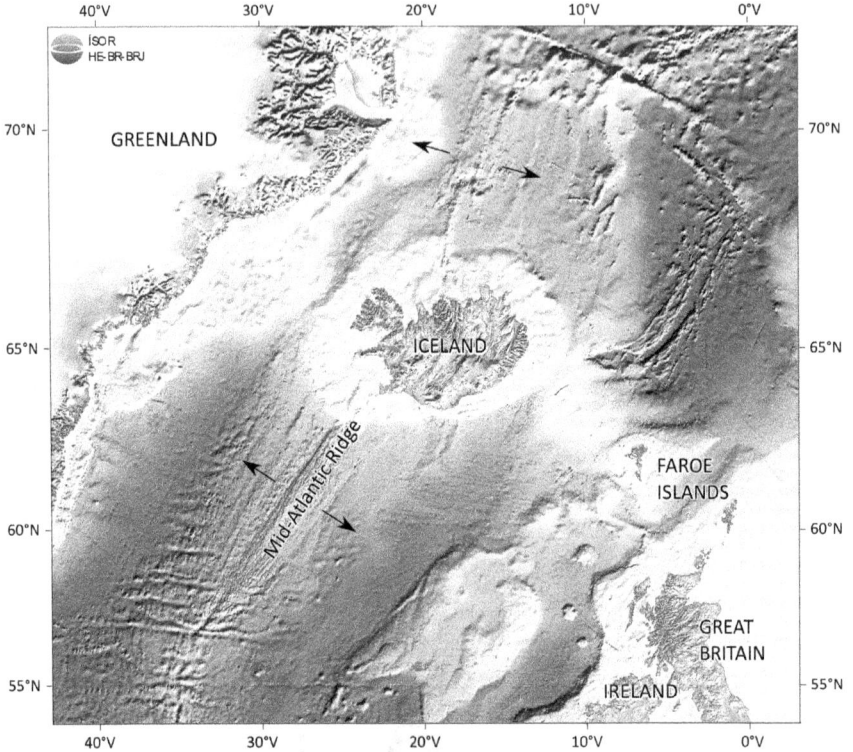

Figure 3.1   The location of Iceland and the rift system of the MAR.

lava that solidifies on the surface moves horizontally away from the spreading axis, with half spreading rate of 1 cm/year, but simultaneously moves vertically down due to the load of lava that accumulates on the surface at a later time. The trajectories of the lava successions are shown in Figure 3.2. The sections that cool at the spreading axis move vertically down with time, while those which accumulate outside the spreading axis move both laterally and downward. When lava has left the volcanic zone, it only moves laterally away from the spreading axis with time.

A consequence of this process is that the lava becomes reheated as it moves to greater depth, pores close due to external pressure and it undergoes hydrothermal alteration, which finally makes the rock almost impermeable. Close to the spreading axis, the subsiding

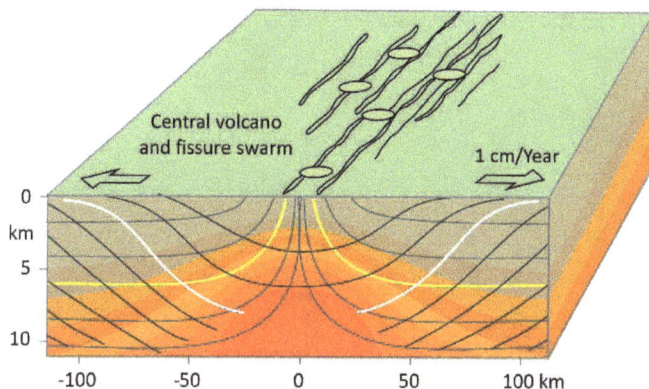

Figure 3.2 Pálmason's crustal accretion model of an active spreading ridge like in Iceland. The yellow and the tiny blue lines show the trajectories of the lava material from the surface away from the rift axis and downward. The white line and the tiny black lines show the isochrones and the temperature is shown with the color scale. As an example a piece of lava that solidifies at the surface close to the rift axis will follow the yellow line with time. During the journey, it is reheated up to a certain maximum due to the downward movement but cools later on as the distance from the spreading axis increases.

crust can even be reheated to solidus of certain minerals so it starts to melt and creates silicic magma (Figure 3.2).

Fresh lava on the surface has very high porosity ($\sim$30%) and primary permeability. Therefore, there is practically no conductive heat transport in the near surface lava pile, all heat from below is removed by groundwater flow. As the lava becomes buried, the porosity and permeability reduce with depth as a consequence of the burial pressure and the precipitation of secondary minerals from geothermal fluids. Temperature logs in boreholes in the volcanic zones in Iceland indicate that primary permeability has been reduced enough at around 1 km depth to let thermal conduction dominate the heat transport. A result of this is that the uppermost 1 km of the volcanic crust in Iceland should have very high permeability. But since repeated glacial erosion has removed the uppermost 1–2 km of the crust at present sea level outside the volcanic zone, the general permeability of the basaltic crust outside the volcanic zone is low.

Heat flow in Iceland is high compared to continental areas (Figure 3.3). The heat flow is basically controlled by two processes.

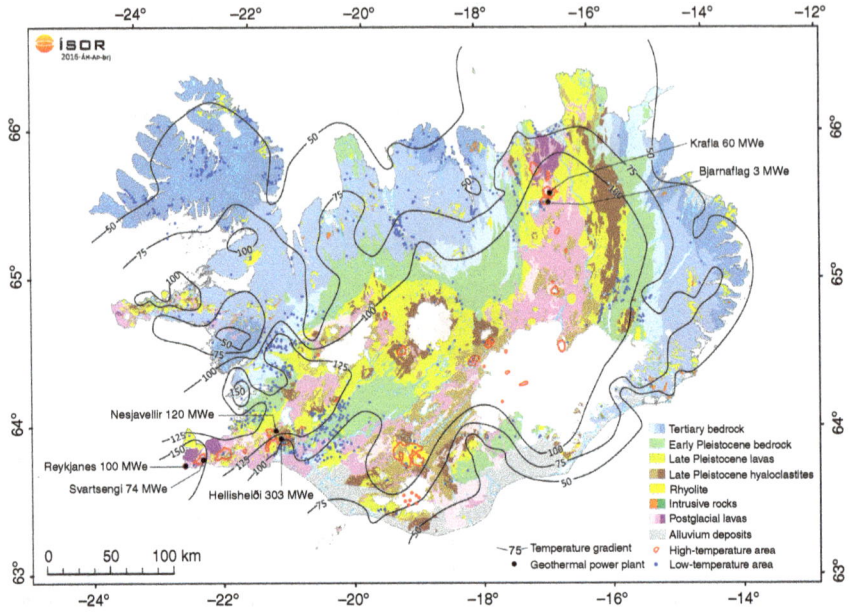

Figure 3.3   Geological map of Iceland, showing high- and low-temperature areas and the isolines of the temperature gradient.

One is a background heat flow, originating from the cooling crust away from the spreading axis like at the mid-oceanic ridges. The other is local high or low heat flow anomalies caused by convection of water in vertical fracture systems, the high values corresponding to the upwelling part, while the low values relate to the down flow pattern. Since the crust in Iceland is fairly homogeneous with respect to thermal conductivity (1.6–1.9 W/m°K), the near surface temperature gradient is frequently used as a proxy for heat flow. The typical background values of the temperature gradient in Iceland is 80–100°C/km at the boarder of the volcanic rift zone lowering to 40–50°C/km in the oldest crust that is farthest away from the rift axis. Within the volcanic zone, however, the uppermost 1 km of the crust consists of highly permeable young volcanics, where all conductive heat from below is transported away by large groundwater currents. Therefore, almost a zero temperature gradient is observed close to the surface within the volcanic rift zone, with the exception of the high-temperature hydrothermal fields associated with the volcanic centers.

3.3.2. Geothermal Fields in Iceland

Geothermal surface manifestations are very common in Iceland. They appear as hot springs, fumaroles, steam vents, or simply as geothermal alteration minerals on the ground. Geothermal areas in Iceland are basically of two different types, high-temperature fields and low-temperature fields, but areas with intermediate reservoir temperatures (150–200°C) are rarely found. The locations of the high- and low-temperature fields in Iceland are shown in Figure 3.3. There are fundamental differences between high- and low-temperature fields.

3.3.2.1. High-temperature fields

The high-temperature fields have reservoir temperatures of 200–340°C and are exclusively located within active volcanoes or recent postglacial volcanism in the axial rift zone. Their surface manifestations are hot springs and fumaroles and high-temperature rock alteration, resulting in colorful and picturesque landscapes (Figure 3.4). The chemical content of the brine from the Reykjanes and Krafla high-temperature fields are listed in Table 3.1. Their geothermal fluid is usually acidic and the rather high chemical content makes the fluid not suitable for direct use.

In the following sections, two different, but typical high-temperature fields in Iceland, are described. These are the Krafla field, which is a representative high-temperature field associated with a central volcano within the volcanic rift zone, and the Reykjanes field, which is located at the tip of the on-land continuation of the rift axis of the MAR (Figure 3.3).

The Krafla high-temperature field is one of the most famous high-temperature fields in Iceland with its 60 MW$_e$ power plant. It has been in operation since 1978, and is owned and operated by Landsvirkjun, the National Power Company in Iceland. Mt. Krafla is located just southeast of the center of a ∼80 km^2 caldera within the volcanic zone of northeast Iceland (Figure 3.5). About 80 km

Figure 3.4   High-temperature area near Mt. Krafla, NE-Iceland. Photo: Sigurður G. Kristinsson.

volcanic fissure swarm crosses the caldera along the active rift zone. The Krafla power plant was the first large geothermal power plant in Iceland to generate electricity. The production drilling and construction of the plant started concurrently in 1975. In December that year, a volcanic fissure eruption began in Krafla less than 3 km away from the construction site. This was the beginning of a nine years long period of volcanic unrest that lasted from 1975 to 1984, and severely disrupted the construction and affected the geothermal reservoir with unforeseen consequences (e.g., Mortensen *et al.*, 2009). The most severe effect was the intrusion of magmatic gasses into a part of the geothermal reservoir, which slowly diminished with time (Ármannsson *et al.*, 1989) but made the fluid very corrosive. The affected part of the system could not be used for production for a long time.

The volcanic activity involved repeated events of continuous uplift caused by accumulation of magma in a shallow magma chamber indicated by seismic observations (Einarsson, 1978). The

Table 3.1 The chemical content of geothermal fluid from few low- and high-temperature fields in Iceland.

	Typical low-temperature fields					High-temperature fields		Standards for drinking water	Typical drinking water in Iceland
	Laugaland	Hveravellir	Kjós	Reykir	Kaldárholt	Krafla	Reykjanes		
Well name	LN-12	HV-01	MV-24	RR-22	KH-36	K-40	RN-12		VK-1
Conductivity (μS/cm)/Temp	264/25	282	387	321	323	—	—	2500	79
pH/°C	9.67/22.7	9.31/21.8	9.58/21.9	9.60/21.5	10.3/21.3	7.68/25	5.50/21.3	6.5–9.5	8.95/25
Carbonate (CO_2) (mg/L)	17.6	33.5	17.5	23.0	18.8	77.3	48.2		—
Hydrogen sulfide, H_2S (mg/L)	0.08	1.45	2.36	1.5	0.12	18.1	1.93		—
Oxygen, O_2 (mg/L)	0.001	0	0	0	0.005	—	—		
Total dissolved solids (mg/L)	248	282	423	295	230	—	38310		
Boron (B) (mg/L)	0.18	0.07	0.14	0.03	0.13	4360	8820	1	<10
Silica (SiO_2) (mg/L)	100.2	169	188	107.9	87.2	461	822		14.5
Sodium (Na) (mg/L)	2.1	53.1	78.3	65.9	64.6	65.9	10800	200	9.69
Potassium (K) (mg/L)	1.09	2.47	3.55	1.79	0.65	12.3	1600	12	0.441
Magnesium (Mg) (mg/L)	0.002	0.002	0.003	0.003	0.002	0.009	1.06	50	0.923
Calcium (Ca) (mg/L)	2.7	1.69	2.55	2.96	2.61	0.94	1930	100	5.27
Fluoride (F) (mg/L)	0.44	0.9	1.99	5.07	2.18	1.86	0.26	1.50	<0.2
Chloride (Cl) (mg/L)	11.9	11.2	21.6	8.17	20.6	21.6	22160	250	7.76
Sulfate (SO_4) (mg/L)	37.8	28.0	53.0	48.8	26.4	13.4	16.1	250	1.82
Radioactivity (Radon) (Bq/L)	1.46	1.01	0.92	5.64	6.75	—	—	100–	—
Reservoir temperature (°C)	90	125	140	75	70	—	300	—	—

(Continued)

Table 3.1   (*Continued*)

	Typical low-temperature fields					High-temperature fields		Standards for drinking water	Typical drinking water in Iceland
	Laugaland	Hveravellir	Kjós	Reykir	Kaldárholt	Krafla	Reykjanes		
Aluminum (μg/L)	180	192	380	21.4	121	1484	85	200	20.7
Arsenic (μg/L)	5.4	1.20	1.43	0	—	24.9	112	10	<0.05
Cadmium (μg/L)	0	0.00	0.014	0.015	—	<0.002	0.055	5	<0.002
Chromium (μg/L)	0.1	0.00	0.302	0.143	—	0.262	0.221	50	0.009
Copper (μg/L)	0.18	<0.1	0	0.102	—	0.142	0.531	2000	0.942
Iron (μg/L)	1.7	3.00	2.32	1.22	4	28	102	200	1
Lead (μg/L)	0.06	0.03	0.012	0	—	<0.01	4.89	10	<0.01
Mercury (μg/L)	0	0.00	0.006	0	—	0.027	0.052	1	<0.002
Manganese (μg/L)	0.2	0.18	0.055	0.249	0	—	3220	50	0.082
Nikkel (μg/L)	0.06	0.16	0.237	0.148	—	0.31	1.66	20	<0.05

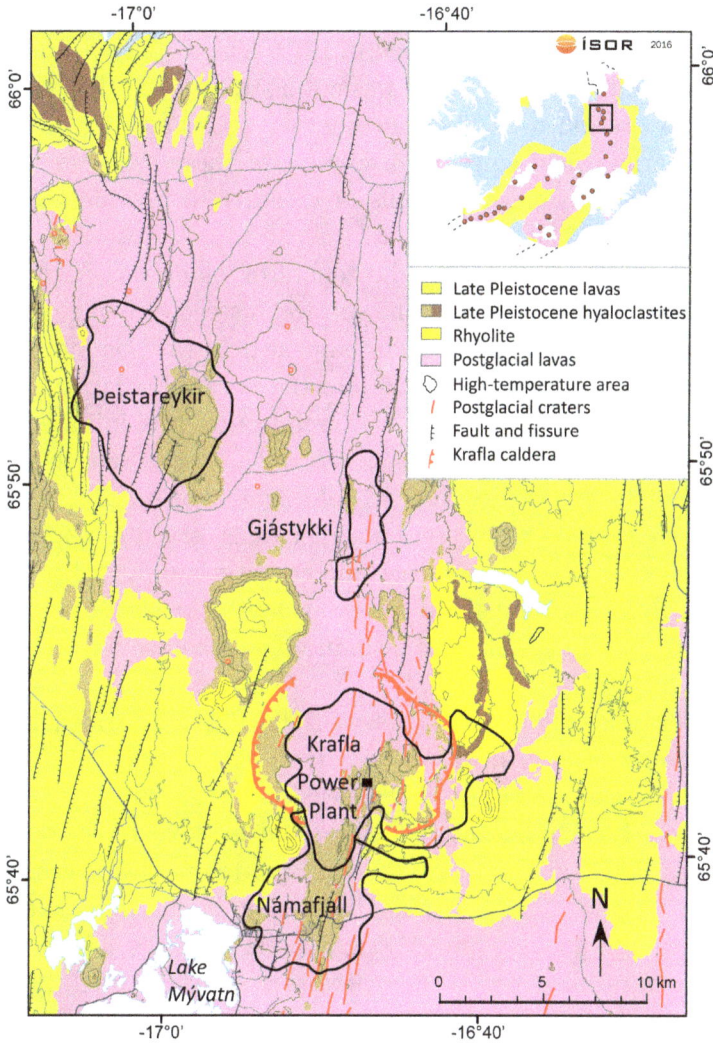

Figure 3.5   The Krafla high-temperature field.

inflation of the magma chamber was regularly interrupted by rapid deflation during short rifting periods lasting several days, where magma was injected laterally into the fissure swarm. On several occasions, this resulted in volcanic fissure eruptions of low viscous basaltic magma that lasted for a few days.

The design of the power plant included two 30 MW_e units, of which the first was commissioned in 1978. Due to the volcanic activity, the second 30 MW_e turbine was not installed until 1996 and came into full operation in 1999. Altogether 44 wells have been drilled in the geothermal field since 1974. Several of these wells were drilled in recent years as make up wells or in order to provide energy for possible enlargement of the plant.

The disposal water from the separator plant was initially discharged into a small pond on the surface, instead of being reinjected into the geothermal reservoir. The effluent from the pond mixed with a small stream that disappeared into the permeable lavas and from there to the groundwater. In 2002, reinjection was initiated and a part of the effluent from the power plant has been reinjected since then. Most of the reinjection has taken place below 2000 m depth through well no. K-26, but several other wells served temporarily for this purpose. On-going microseismic activity is reported in Krafla, mostly within the wellfield at reservoir depth. At least part of these small earthquakes are likely to be induced by the injection (Ágústsson et al., 2012).

In the years 2008–2009, the well IDDP-1 was drilled into the Krafla geothermal field. It was a part of the Iceland Deep Drilling Project (Pálsson et al., 2014; Fridleifsson et al., 2005). The target was supercritical fluid that was expected at 4500 m depth, which is proposed to exist at the roots of high-temperature geothermal areas in Iceland. The well was sited using the existing knowledge of the geology and the underlying magma chamber and partly on the basis of results from 1D interpretation of magneto-telluric (MT) and transient electromagnetic measurements (TEM) resistivity data as well as depth distribution of local earthquakes and the results of drilling of a nearby well. The concept was to drill and case the well with cemented steel casing down to 3500 m depth, which was the top of the target zone, and to install a slotted liner in the production part below that depth.

Surprisingly, the well entered an 800–950°C rhyolitic magma body, close to 2096 m depth (Schiffman et al., 2014) and further drilling was terminated. A high permeability layer of 30–40 m lies above the magma body generating 450°C of superheated steam, which was discharged into the atmosphere during testing. It was estimated that the well could yield 25–40 MW of electricity depending on the turbine configuration (Ingason *et al.*, 2014) providing that issues with the chemistry of the fluid could be solved. After almost 28 months of flow testing of IDDP-1 and experiments to harness the well for power production, a failure in the wellhead equipment led to the necessity to kill the well with water injection (Einarsson *et al.*, 2015). After being used for reinjection for a while an attempt was made to clean the well. However, serious casing damages were observed in the well, the triple casing was pulled apart and the well could not be secured for blow-out, and was therefore cemented to the bottom and closed permanently. Despite this borehole failure, the experience from IDDP-1 indicates that there are possibilities to extract large amounts of geothermal steam from the magma contact zone within active magmatic geothermal systems. Harnessing this fluid needs further research and experiments on handling chemical problems as well as borehole design and material selection for well casing and wellhead equipment.

The Reykjanes high-temperature field is located at the rift axis of the Reykjanes Ridge where it emerges from the sea into Iceland (Figures 3.3 and 3.6). It is characterized by an active fissure swarm that curves into the Reykjanes transform zone where the active spreading axis is shifted eastward by about 60 km through a series of parallel en-echelon rift systems (Figure 3.6). The actual geothermal field is located on the axis close to the coast and is characterized by around 2 km^2 of intensive hydrothermal activity and geothermal alteration at the surface. The hydrothermal activity has been highly variable with time. It increases normally as a consequence of large earthquake episodes that occur with intervals of a few decades. The largest reported earthquakes are estimated to be close to magnitude 5.1.

The Reykjanes high-temperature field differs from other high-temperature fields in Iceland in the chemical content of the fluid, being of seawater origin (Table 3.1). Limited development of the geothermal field originally started in 1956 with shallow wells, and the

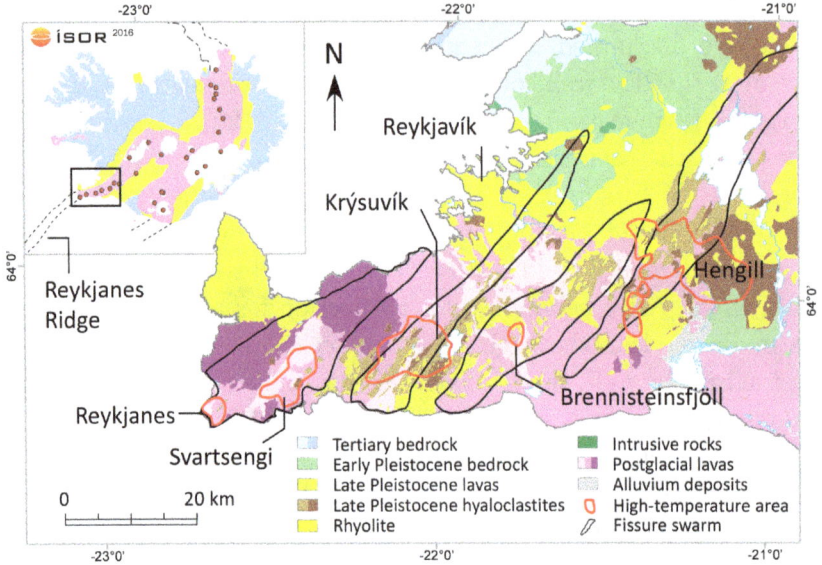

Figure 3.6   The Reykjanes peninsula showing the Reykjanes high-temperature system. The en-echelon fissure swarms of the peninsula are also shown and the high-temperature system located within them.

field was utilized for intermittent and experimental industrial applications, chemical processing, and small-scale electricity generation for plant use in the latter half of the 20th century. Altogether nine wells were drilled during this period. The high chemical content of the fluid created interest in extracting chemicals from the geothermal brine and a geothermal salt factory was operated in the latter half of the 20th century, but it was not economical in the long term and so was abandoned. A fish drying plant was also operated there using the geothermal steam. By the end of the 20th century, the company Hitaveita Suðurnesja, later HS Orka, planned a large-scale electricity production at Reykjanes, and in 2006 a new 100MW$_e$ power plant came into operation. The plant is composed of two 50MW$_e$ condensing turbines where 4000 L/s of cold seawater is used to condense the steam. The reservoir temperature is 290–320°C and the wells are typically up to 2500 m deep, yielding a two-phase flow in the boreholes. A steam cap has evolved at shallower depths due to pressure decline in the reservoir and is used to produce pure steam from shallower wells. Due to the high chemical content of the reservoir fluid the wellhead pressure is kept high, at 24 bar, and the steam separators are operated at 18 bar to reduce scaling in the wells. The most difficult scaling are the sulfides that form at a relatively high pressure, while amorphous silicates form at lower downstream pressures. The separated brine has to be diluted with condensate prior to reinjection to prevent scaling in the reinjection wells. Despite these measures, scaling must be cleaned regularly, both in wells and in surface installations. Altogether 34 wells have been drilled into the geothermal field until 2016, of which 24 were drilled for the power plant development.

Analysis of cuttings and well loggings from wells in Reykjanes show that the stratigraphy consists of basaltic formation interbedded with shallow marine fossiliferous sediments (Friðleifsson *et al.*, 2014). The basaltic formations are mainly submarine hyaloclastite formations with increasing proportion of basaltic lavas and intrusions at deeper levels.

3.3.2.2. The low-temperature fields

The low-temperature fields in Iceland show quite a different character from those in the high-temperature fields. Their location is shown in Figure 3.3. The low-temperature geothermal systems are almost all outside the axial rift zone. Their reservoirs are fracture dominated in otherwise low-permeability basaltic lavas or hyaloclastites. The heat is extracted from the relatively high background temperature gradient by fluid convection in permeable fracture systems (Flóvenz and Sæmundsson, 1993). In most cases, the surface expressions of the low-temperature fields appear as hot springs on the surface (Figure 3.7), frequently as linear rows of hot springs indicating the underlying fractures that bring the hot fluid finally the surface. Sometimes, the low-temperature fields are hidden with no expression of geothermal activity at the surface (Axelsson *et al.*, 2005). In these cases, the geothermal fields are discovered by geophysical measurements of heat flow, or electrical resistivity of the subsurface, or simply by tectonic data.

Figure 3.7   Krosslaug, a hot spring in West Iceland, about 40°C warm. Photo: Sigurður S. Jónsson.

The tectonic origin of the permeable fractures is not always obvious. In some cases, they are clearly linked to the shear zones of the transform faults between rift segments, like in the South Iceland Seismic Zone. There are also examples of low-temperature fields where recent or active fissure swarms of the volcanic centers penetrate the older rocks without any surface volcanism. It has also been suggested that the geothermal fractures are a consequence of the postglacial rebound of the crust (Bödvarsson, 1982). As expected for a region where the MAR interferes with a hot spot, the overall seismicity in Iceland is high (Jakobsdóttir, 2008). Figure 3.8 shows a map of the seismicity of Iceland for the period 1995–2016, embedded on a simplified geological map. The seismicity shows clearly the main tectonic and geological patterns of the country; the volcanic centers in the axial rift zones, the axial part of the MAR, and the transform faults and oblique rift zones in North and South Iceland, which connect the rift segments in Iceland. The eastern part of Iceland outside axial rift zone, i.e., that belonging to the European plate, is almost

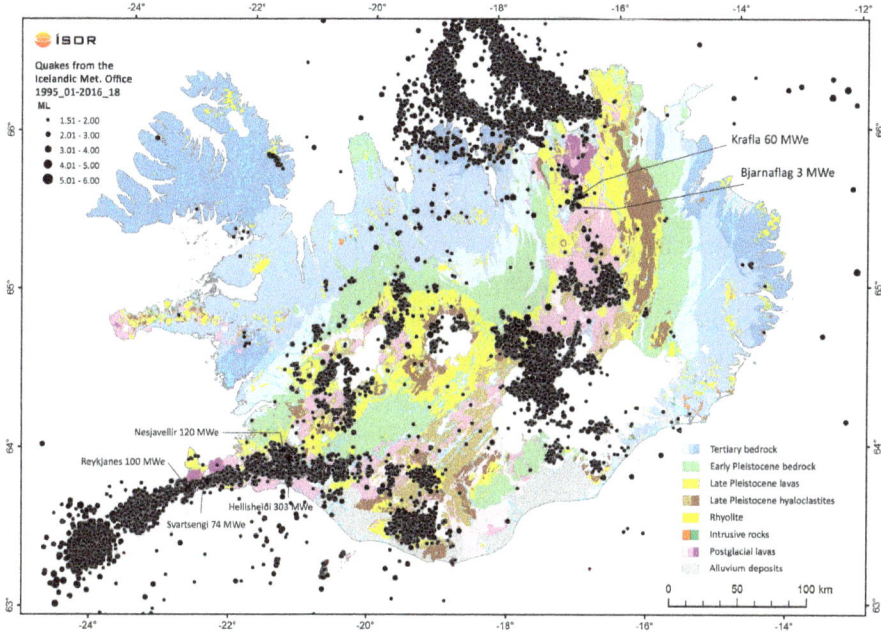

Figure 3.8 Seismicity of Iceland for the period 1995 to early 2016. The data are from the SIL network catalog, the national seismic network operated by the Icelandic Meteorological Office (IMO). Only events with magnitude higher than 1.5 are shown. Location of the geothermal power plant is also shown.

devoid of seismicity as well as surface geothermal activity. On the contrary, the part that belongs to the American plate outside the axial rift zone is characterized by both intraplate earthquakes and hydrothermal systems supporting the hypothesis of tectonic origin of the low-temperature systems.

The chemical content of the geothermal fluid within the low-temperature system is usually quite low, typically with total dissolved solids of less than 300 mg/L. At a few places, the reservoir fluid is slightly contaminated seawater, giving rise to total dissolved solids of over 1,000 mg/L. As an example, the chemical content of the fluid from three low-temperature fields are listed in Table 3.1. The quality of the geothermal water is indeed mostly within drinking water standards. The fluid is eminently suitable for domestic use and provides tap water and water for radiator systems.

3.3.3. Resource Estimates

In 1985, Pálmason *et al.* published a comprehensive geothermal assessment of Iceland. It was a volumetric assessment of the total energy stored in the crust beneath Iceland based on knowledge at that time of crustal temperature distribution. The thermal energy stored in the uppermost 3 km was considered to be recoverable with current geothermal drilling technology in 1985. Only a small proportion of the available energy was believed to be accessible due to geographic constraints (below high mountains and glaciers) and due to limits of the possible maximum thermal recovery that was estimated to be from 1% to 20% depending on geological conditions. Applying these constraints to the calculated available energy led to the assessment of technically exploitable thermal energy. Further losses of 92–98% were accounted for during transformation of the thermal energy from the borehole fluid into electricity, depending on the reservoir temperature. Note also that the lower temperature limit of the thermal energy for power production is considered to be 130°C. Nowadays, it would be more reasonable to use 80°C as the lower limit, assuming the use of binary systems.

In the assessment calculation, the country was divided into the volcanic zone and the nonvolcanic zone. The former was also subdivided into active areas and nonactive areas. The active areas consist of the active volcanic systems, i.e., the central volcanoes and their fissure swarms. Within the active areas, there are about 30 identified high-temperature hydrothermal fields similar to those shown in Figure 3.3 and they were treated separately.

The result of the assessment from 1985 was that nearly 19,000 TWh of electricity from geothermal energy could be produced in Iceland, which is equal to 43,000 MWe for 50 years. Only a small part of this (1600 TWh or 3500 MWe for 50 years) is within the known high-temperature geothermal fields. Hence, by far the majority of the potential for geothermal electricity generation in Iceland is outside the presently known high-temperature geothermal fields, but mostly within the volcanic rift zone. It is important to note here that these numbers are only an estimate of the technical harnessable energy and do not take economical or environmental aspects into consideration.

Since 1985, several things have changed. First, several high-temperature fields have been drilled and some utilized for electricity and heating production. In some cases, they have proven to have considerably more generating capacity than indicated by the volumetric assessment, but others have turned out to be less productive than expected. The volumetric assessment of 18 known high-temperature fields has therefore been revised (Ketilsson *et al.*, 2009) and the resulting number is now 4300 MW for 50 years instead of 3300 MW, a 30% increase.

The assessment from 1985 assumes the lower limit of accessible geothermal energy at a 3 km depth. Since then, the drilling technology has advanced significantly and today it is more reasonable to assume the lower limits to be at 6 km depth. This will increase the accessible geothermal energy beneath Iceland several times due to the high temperatures between 3 and 6 km. The background temperature gradients observed in Iceland are of the order of 40–140°C/km outside the volcanic rift zone; inside they might be still higher below the permeable layer. This means that temperature exceeding 200°C can be expected at 2–4 km depth almost everywhere beneath Iceland, since available data indicate that background geothermal gradients can be extrapolated almost linearly down to temperatures of 600–700°C (Ágústsson and Flóvenz, 2005). Interpretation of seismic and temperature data, as well as laboratory data, also show that the brittle ductile boundary in the basaltic crust of Iceland is close to the 700°C isothermal surface (Ágústsson and Flóvenz, 2005; Violay *et al.*, 2009). This boundary can be expected to be the absolute lower boundary of accessible geothermal energy. Therefore, it is suggested that the region at below the volcanic zone of Iceland (32,000 km^2) has temperatures of 250–300°C at 3 km depth increasing to 400–600°C at 5 km depth. The challenge of the near future is to prove or disprove this hypothesis by more accurate and sophisticated geophysical measurements, supported by laboratory experiments and followed up by deep drilling. The result will be an important input to a revised geothermal assessment of Iceland. In addition to the general hot upper crust of Iceland, considerable energy is contained in volumes of partially molten magma that exists at 3–6 km depth beneath several of the central volcanoes.

3.4. Sustainable Energy Production

The long experience of geothermal energy production in Iceland has revealed the need to manage the energy production of the geothermal resources in a sustainable way. To do so, the first step is to define the concept of sustainable energy production. A group of Icelandic geothermal experts came up with the following definition that has proven to be quite useful

For each geothermal system, and for each mode of production, there exists a certain level of maximum energy production, E_0, below which it will be possible to maintain constant energy production from the system for a very long time (100–300 years). If the production rate is greater than E_0, it cannot be maintained for this length of time. Geothermal energy production below, or equal to E_0, is termed sustainable production while production greater than E_0 is termed excessive production (Axelsson *et al.*, 2001).

This definition requires that in order to be sustainable each geothermal field must be possible to maintain nearly constant energy production from the resource over one to three centuries, which is five- to 10-fold the typical economical depreciation time of a power project. It also implies that the level of sustainable energy production is technology dependent and can, for example, be higher if reinjection is applied. Furthermore, the level of sustainable production is never well known before the start of the production but will appear slowly during the production phase. In the beginning of utilization, the production is in most cases excessive but levels off with time toward the sustainable level. If it does not level off, the production must be reduced of its mode changed until the appropriate sustainable level is reached. It should also be noted here that the definition does only apply to the production itself but not the environmental or social impact of the production.

The definition of the level of sustainable energy production does not exclude the possibility of periodic variation in production to

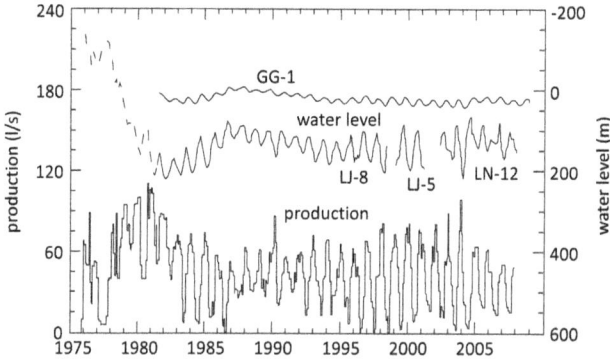

Figure 3.9   Production and water level in the years 1975–2008 of the Laugaland low-temperature field in northern Iceland. The water level has been monitored during the operation time in several boreholes like GG-1, LJ-8, LJ-5, and LN-12 (figure from Axelsson *et al.*, 2010).

adapt the production to the energy market at each time. Thus excessive production interrupted with periods of low production can be regarded as sustainable as long as the long-term average or the production is sustainable (Axelsson, 2010). Seasonal variation in production is a classic example of periodic excessive production but the period could also be considerably longer. This is demonstrated in Figure 3.9 that shows the production and water level during 38 years of production from the Laugaland low-temperature field in northern Iceland, which is exploited for district heating (DH) (Axelsson *et al.*, 2010). It shows the typical seasonal variation in production for a district heating system but also a clear excessive production during the first five years, which had to be addressed by strong reduction in production until sustainable level of production was reached and maintained since.

3.5. Environmental Impact of the Geothermal Production

The main environmental impact of the geothermal production in Iceland is the positive influence on the quality of the atmosphere with reduction in carbon dioxide emission and other pollutants from burning of fossil fuel. The production of hot water from low-temperature fields is practically without any emission of gasses. Compared to the alternative burning of fossil fuel for heating purposes, the annual CO_2 savings by using the geothermal energy for district heating amounts to 2–4 million tons of CO_2 annually (Gunnlaugsson *et al.*, 2015). A slight content of hydrogen sulfide (H_2S) is usually observed in the low-temperature geothermal water and gives a faint smell when the hot water is used directly for bathing and washing. The inhabitants are used to this smell and do not even notice it on daily basis but foreign guests may notice it. The concentration of H_2S is far below all health limits but our nose detects the H_2S in extremely low concentrations. In case of oxygen contamination of the hot water, the H_2S in it reacts with the oxygen to produce sulfate and has therefore a corrosion-preventing effect. Therefore, H_2S is sometimes deliberately mixed into the hot water if its concentration is negligible. Otherwise the environmental impact of energy production from the low-temperature fields is negligible, no environmentally harmful chemicals follows, small land use for installations, and most of the pipes are underground. The natural hot springs will, however, disappear as a consequence of pumping water from geothermal wells but they appear again if the production is stopped for some time.

In case of the high-temperature fields, the environmental impact of the production is considerable higher. The impact is mainly due to emission of CO_2 and H_2S, visual effects, induced seismicity, and disposal of the effluent from the power plant. Contrary to the low-temperature fields, the production of energy from the high-temperature fields rather increase than decrease the geothermal activity at the surface.

Emission of CO_2 and H_2S follows the geothermal steam but in different concentrations from one field to another. In case of Icelandic power plants, the emission of CO_2 per electricity produced is in the range 26–181 g/kWh$_e$ (Ármannsson *et al.*, 2005; Sigurdardóttir and Thorgeirsson, 2016). This is of the order of 4–25% of similar values for oil-driven power plants. Although the concentration of H_2S from the power plants in the atmosphere is far below environmental limits, it may be quite annoying and cause damages to electronic equipment and corrode metals like copper and silver. In case of, the 303 MW$_e$ Hellisheiði power plant, the largest geothermal power plant in Iceland, ongoing research and demonstration projects called Carbfix and Sulfix show that it is possible to collect and reinject the CO_2 and H_2S into the shallower part of the reservoir where it fixes permanently in minerals as calcite and pyrite. In 2015, over 10% of the emitted CO_2 was reinjected and research indicates that at least 90% of it is fixed in mineral form within a year from reinjection. At the Hellisheidi power plant, about 30% of the H_2S was reinjected in the Sulfix project in 2015 whereof about 75–80% is sequestered in the form of minerals like pyrite within six months (Sigurdardottir and Thorgeirsson, 2016). These leading experiments of Reykjavik Energy are paving the way for future zero emission geothermal power plants. As explained later in this chapter, alternative solutions to deal with the CO_2 emission is to use it for production of synthetic fuel or to fix it in plants in greenhouses.

Induced seismicity is commonly a concern for geothermal energy production in the world, especially in the case of hydraulic stimulation. In general, this has not been the case for Iceland with one exception. As Figure 3.8 shows, earthquakes are quite common in Iceland and people living close to the plate boundary and geothermal fields are used to small earthquakes now and then. Therefore, it was generally considered that induced seismicity would just occur as small additional noise on the background seismicity and not disturb the public. Monitoring of earthquake shows that production-induced earthquakes are common at all the high-temperature fields but are not observed at the low-temperature fields unless they are located directly at the plate boundary (Flóvenz *et al.*, 2015). The magnitudes of the induced earthquakes are normally less than 2.0. However, historical natural earthquakes of magnitude up to 6.6 are known to occur in some of the geothermal fields but these events will happen anyway regardless of the geothermal energy production. In 2011, large-scale reinjection was initiated at the Hellisheiði power plant, which triggered repeated swarms of earthquakes up to magnitude 3.9 (Bessason *et al.*, 2012), which were strongly felt in the neighboring village and annoyed people severely. Based on this experience, the authorities have introduced guidelines for reinjection operations in line with what is being done in the international community.

In high-temperature fields in Iceland, the production wells usually yield a mixture of water and steam, which are separated using steam separators. The steam is used to run the turbines while the effluent from the separators needs to be disposed. The effluent is a geothermal brine with high concentration of silica and other chemical substances. The brine was initially disposed at the surface or into the shallow ground water while later, it has been reinjected back into the geothermal reservoir. In the case of the Svartsengi power plant, the effluent was disposed on the surface, which led to formation of a blue-colored pond outside the power plant. Later, people stared to have bath in the pond and it became more and more popular with time and bathing in the pond was deemed to have skin healing effects. This pond that originally was an industrial waste is now the world famous Blue Lagoon in Iceland attended by hundreds of thousands every year.

The most debated environmental impact of the geothermal power plants in Iceland is the visual impact of buildings, boreholes, power lines, roads, and drill sites. The reason is that the high-temperature fields are mostly located in areas of unique geology, often in the highlands with spectacular nature and colorful landscape. The debate is about to what extent these areas should be protected and used only for tourism or power production should be allowed with restrictions.

In 1999, the government of Iceland launched a project to create a master plan for environmental protection and sustainable utilization of energy resources of the country. The purpose was to evaluate potential energy projects in hydro power and geothermal and rank them with respect to their energy and economic potential as well as the estimated environmental and social impacts. The master plan should rank the suggested power projects into three groups with respect to these factors, a group of sites or projects that are suggested to be used for energy production, a group of sites that should be protected, and a group of projects where further information is required for decision. The outcome of the master plan is then subject to parliamentary approval. It is a living document that is continuously revised and discussed but the Icelandic parliament has the final decision power. The master plan is also a helpful tool to assist decisions makers to filter out which projects are likely to become controversial and disputed and which ones are not. It also directs the attention to those project areas that might have protective value and should be left untouched by human development.

3.6. Geothermal Utilization in Iceland

Iceland was colonized in the last decades of the 9th century, when settlers came from regions where volcanism, earthquakes, hot springs, and geysers were unknown phenomena. There is, however, surprisingly scant documentation of geothermal activity in the ancient literature of Iceland. According to stories, the hot springs were used as social meeting places where people would share a common bath. Due to a lack of technology and economic resources, it was not until the beginning of the 20th century that Icelanders began to make use of both geothermal and hydropower. To begin with, utilization was on a small scale but this rapidly increased as the century progressed. Geothermal is now the largest source of primary energy in Iceland (Figure 3.10), at about 68% in 2014 (Ragnarsson, 2015), higher than anywhere else in the world.

Icelandic geothermal resources have been used for both electricity generation and direct use such as space heating, horticultural food production, swimming and bathing, fish farming, the processing of salt, and other industrial uses (Figure 3.11). Space heating is by far the most important direct utilization, with 90% of all space heating in Iceland coming from geothermal.

The engineer and pioneer Baldur Líndal introduced, in 1973, a diagram for utilization of geothermal water and steam suitable for various applications. A modified version of his diagram is shown in Figure 3.12. The hot fluid can be exploited through a series of utilization processes, which require lower and lower temperatures (Pálmason, 2005). Several Icelandic firms employ Lindal's method, making use of excess energy from other companies. What is waste for one can be valuable for another.

PJ

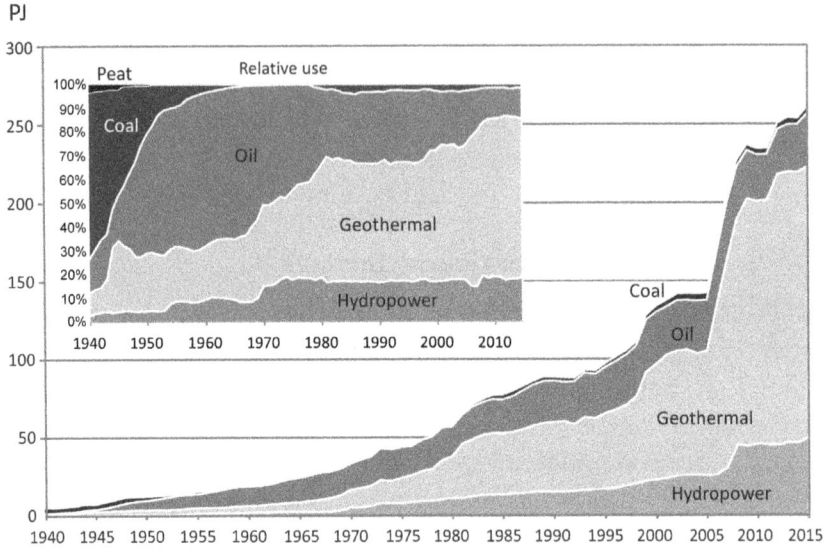

Figure 3.10   Primary energy use in Iceland from 1940 to 2015 (Orkustofnun, 2016).

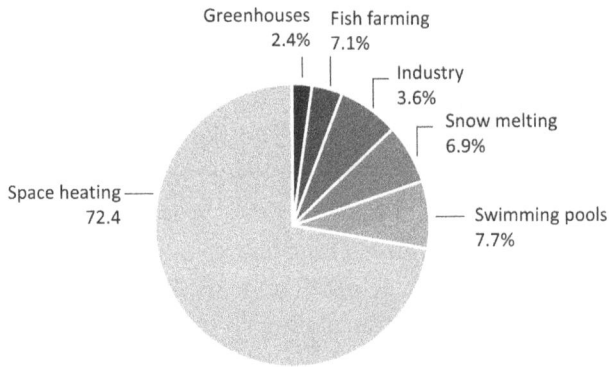

Figure 3.11   Utilization of geothermal energy in Iceland categorized by the International Geothermal Association classification (Orkustofnun, 2016).

3.6.1. Space Heating

When it is cold outside or unpleasant weather, it is always enjoyable to come home into a warm building. To most Icelanders, this is not a luxury. It is taken for granted that hot water of 50–80°C will flow through their heating systems all day and all year around, during

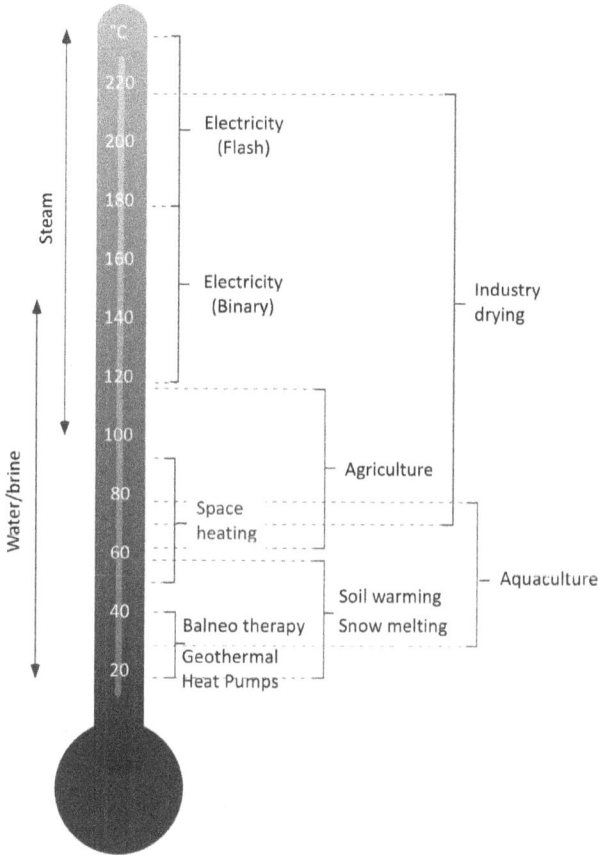

Figure 3.12   Utilization of geothermal water and steam (modified from Lindal diagram).

summer and winter. The home owner controls the room temperature at around 22°C. For safety reasons, some people may have heat exchangers to lower the temperature of the tap water. Foreign visitors may notice that Icelanders are profligate with energy use. People tend to open their windows, or even the front door of their home, to reduce the indoor temperature. There is always hot water available for washing and bathing. It is because of this access to relatively cheap and environmentally friendly geothermal water that Iceland differs from other countries in geothermal utilization.

More than a century ago, major advances in the utilization of low-temperature water took place when plumbing was introduced, originally in Idaho in the United States in 1900 (Pálmason, 2005). At same time, Icelandic farmers made the first attempts to heat their homes using geothermal fluid.

In 1908, Stefán B. Jónsson, a farmer at Suður Reykir in county Mosfellssveit, close to the capital Reykjavík, was the first to channel warm water from a nearby spring, 450 m to his home. Canvas and hemp were wrapped around the pipes for insulation. The water was then passed through radiators in all rooms on the lower floor and produced sufficient heating both for his home and cowshed. According to Stefán, the room temperature was 16°C on a cold winter day when the outside temperature was −17°C. He promoted the idea that Reykjavík should follow his lead and utilize the springs in the warm valley to create a district heating system.

In 1911, another farmer, Erlendur Gunnarsson at Sturlureykir in Borgarfjörður in SW-Iceland, designed the first steam distribution system in the country. He harnessed the power of a hot spring, which was situated 6 m lower than the farm. The hot spring was covered and steam was channeled in a pipe into the house. Later, a stove was connected to the steam supply. This enabled the family to prepare food using steam rather than cooking over an open fire (Þórðarsson, 1998).

Those first attempts proved to people that utilization of geothermal water was possible and the idea spread widely. At this same time early in the 20th century, when villages were developing around Iceland and imported coal was the main fuel, there was often a problem of fuel shortage in rural areas. Interest in the harnessing of geothermal energy grew in proportion to the price of coal. With the new awareness that geothermal water for space heating was economically viable, the decision was taken to build boarding schools all over the country, with the first being at Laugar in northern Iceland (Þórðarsson, 1998).

Utilization on a larger scale began when the Reykjavík Electrical Company drilled for hot water in Laugardalur east of the capital in 1928. In the following two years, 14 wells, 20–246 m deep, were drilled. The original idea was to harness steam for electricity production. However, the resulting 14 L/s of 87°C hot water was not enough to produce electricity, but ideal for space heating. The first buildings to connect to a 3 km long hot water pipeline, were a school, the National Hospital, Reykjavík Swimming Hall, and 60 other houses in the neighborhood. After this great success, it was clear this was only the beginning. It was also clear that this first district heating system could not supply enough water. In 1933, around 3% of homes in Reykjavík were heated with geothermal water. Other geothermal fields, 18 km from the city, were being considered and an agreement was reached to drill near the farms of Reykir and Reykjahvoll. The first eight holes produced 57 L/s of 84–86°C hot artesian water. The district heating system was expanded gradually, step by step, over the years to all of the greater Reykjavik area.

The first district heating system outside Reykjavik was established in 1944 in Ólafsfjörður in northern Iceland (Pálmason, 2005). The oil crisis of 1970 lead to the biggest leap forward that was taken in the expansion of the district heating system. Local and national governments formulated a policy to promote geothermal energy for space heating. These were targeted actions to meet the threat of the oil crisis and yielded lasting results. As can be seen in Figure 3.13, the share of geothermal energy increased from 43% in 1970 to 80% in 1982, and finally reached the current level of 90% in 2010. It took only 12 years to dispense with the need for oil in space heating.

At present, about 30 separate geothermal district heating systems are operated in towns and villages around the country, and in addition there are some 200 small systems in rural areas. These smaller systems supply hot water to individual farms, or groups of farms, as well as holiday houses, greenhouses, and other users (Ragnarsson, 2015).

The application of the hot water does not end here; the spent water from space heating systems, which is about 35°C, is often used to heat pavements and parking spaces. At times, when hot water usage is high, larger systems can also mix the returned waste water with the hot water of 80°C from the direct district heating system to maximize the supply. Householders often use the excess water for their own private hot tub in the garden. Furthermore, that water may

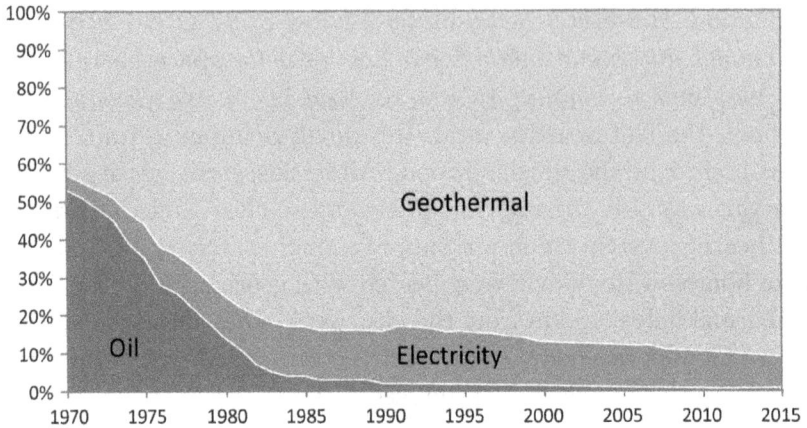

Figure 3.13   Space heating 1970–2015 (Orkustofnun 2016).

be used to melt snow. The first snow melting system in Iceland was incorporated into the steps in front of Reykjavík Junior College in 1952. Today, almost all new buildings have geothermal snow melting systems, which, in terms of piping, cover an area of about 1,200,000 m², mostly in the capital city (Ragnarsson, 2015).

Without doubt, the vision of the farmer at Syðri Reykir, who in 1908 recognized the potential value of the geothermal resource, has been realized. However, it is unlikely that he understood the enormous benefits that would accrue to the country, both economically and in terms of improved living standards for the population.

3.6.2. Food Production

Traditionally, agriculture and fishing were the main industries in Iceland and basic diet consisted of fish and livestock. The cuisine was mainly shaped by natural conditions. However, it was necessary to enhance the diet with food imports, especially with cereals.

Today people enjoy an abundance of food, which they can choose to buy ready prepared, or even visit restaurants. At the same time, people are more interested in the provenance of the food. Culinary tourism is becoming popular. Local and traditional cuisine is sought after. Many restaurants and even greenhouses have capitalized on this, welcoming tourists to their tables. What used to be hearsay or found in ancient travelogues, such as cooking food by using hot springs or baking bread in hot soil, is now a business concept. Soil cooking has reappeared and some restaurants are offering guests food cooked with geothermal steam or baked in hot soil. Iceland has a great deal to offer as a unique place for food lovers. Because of its geographical location and isolation in the North Atlantic Ocean, Icelandic agriculture is almost free of infectious diseases. It has abundant natural resources including pure groundwater, fresh air, and geothermal energy.

There have been many attempts made over several decades to utilize geothermal energy for food production. Cheese and fruit production proved to be unsuccessful but horticulture and vegetable farming is one of the oldest and most widely practiced form of geothermal utilization after space heating.

In the following paragraphs, the utilization of geothermal energy for food production will be demonstrated by describing a few industries and services that base their activity on geothermal energy. This is far from being a complete coverage of the geothermal in the food industry but it should give the reader a taste of the variety of applications of geothermal energy in Iceland.

3.6.2.1. Greenhouses

In the latter half of the 19th century, shortly after potatoes were imported to Iceland, farmers realized that where the soil was warmer it would be better suited for farming. Cultivation near warm springs was tried and was successful. Half a century later, vegetable farming was underway in greenhouses using geothermal as the energy source.

Geothermal heating of greenhouses began in 1924 when the first greenhouse was built at the farm Suður Reykir, SW-Iceland. At first, the water flowed through pipes, which soon got clogged because of mineral residue in the geothermal water. Consequently, geothermal water was used to heat freshwater.

Today the majority of greenhouses in Iceland are located in the South, close to low-temperature resources. A variety of vegetables is grown, such as tomatoes, cucumbers, peppers, salad crops, strawberries, herbs, and also flowers.

Geothermal energy is used in various ways in Icelandic horticulture including:

- To heat greenhouses and to sterilize the soil.
- To produce electricity to illuminate the plants in order to extend the growing season throughout the year.
- To provide pure carbon dioxide to increase the photosynthesis (Pálmason, 2005).

Most greenhouses in Iceland are high-tech, with automatic systems controlling the indoor climate, temperature, lighting, the heating, and the opening of the windows.

Friðheimar is an example of a greenhouse farm in South Iceland. It is a family-run business, which has combined horticulture and tourism (Figure 3.14). Tomatoes are grown all year round and the farm is open to visitors who may tour the greenhouses and enjoy meals in the family restaurant, where they can sample a taste of the crop served among the tomato plants. Tourists are given an excellent opportunity to follow the production method and may also buy freshly grown vegetables and tomato products from the on-site shop. The temperature of the water, which is used in the greenhouses, is 95°C and about 100,000 tons are used per year.

The greenhouse farm is digitally controlled with computers varying the heat, humidity, carbon dioxide, and lighting. The computers are linked to fertilizer blenders, which water the plants according to a predetermined schedule. Weather stations on the rooftops obtain information on wind speed, wind direction, temperature, and sunlight. All these devices are, in turn, connected to the main computer, which is linked to the Internet. This allows the operating managers at Friðheimar to go online, wherever they are in the world and monitor status, change settings, and control watering (Fridheimar, 2015).

3.6.2.2. Mushrooms

The cultivation of edible mushrooms is a specialized branch of horticulture and differs from other greenhouse plants since there is no necessity for artificial lighting for growth. Mushrooms obtain all their nutrition from the decomposing residues of other plant organisms.

Figure 3.14   Inside one of the greenhouses at Friðheimar, where tomatoes and tourists are together (Photo: Brynja Jónsdóttir).

An important part of mushroom cultivation is the production and preparation of the compost. For this process, geothermal energy is used to disinfect raw material in the compost. Water at 70°C is used to sterilize the room for production and to clean the soil before each cultivation. Mushrooms have been grown commercially in Iceland since 1960. A milestone was reached in 1984 when work began on large-scale cultivation at Flúðir in South Iceland. This is the only mushroom producer today and has achieved a dominant position in the home market due to, amongst other things, geothermal energy.

3.6.2.3. Fish industry

Fishing has always been an integral part of Icelandic culture and the fishing sector has been one of the cornerstones of the Icelandic economy. Icelanders have also been in the forefront of production and development of new fishing gear and techniques for the industry.

The use of geothermal energy in seafood processing is mostly in drying and aquaculture. Traditionally, fish was dried in order to preserve it. The fish, usually cod, was hung on wooden stock racks in the open air, and as such called stockfish. Indoor drying began about 40 years ago. Geothermal energy is used to dry the fish, which enables the production all year round. Geothermal heated air is blown over the fish, thus drawing out the moisture from the raw material. There are several fish-drying companies in Iceland. The main market for dried fish is in Africa, particularly Nigeria (Arason, 2001).

Haustak, which was established in 1999, is the largest producer of drying fish products in Iceland. They use high-temperature steam from the power plant at Reykjanes for their indoor drying process. The whole process takes about 2–3 weeks. Haustak dries almost every part of the fish and turns it into a dried food product, including stockfish, heads, bones, and slices. Other fish products include dried fish skin and whole fillets. The original water content in the stockfish is typically reduced from around 80 to 12–14% by drying, but all its nutritional value and taste remain the same. The location at Reykjanes is favorable, being close to the vital resources, both fish and geothermal energy.

The growth of aquaculture has been slow in Iceland. The first experiments date back to 1884, but there have been ups and downs over the years mainly due to weather conditions. The first trials with cage aquaculture on land were made early in the 1970s. This proved to be an unsuccessful method. Open air coastal fish farming suits the weather conditions. There are about 70 fish farms in Iceland and between 15 and 20 utilize geothermal water. The dominant species are salmon and arctic char followed by trout (Ragnarsson, 2015). The advantage of using geothermal water in aquaculture is to shorten the growth period and so increase productivity. This applies to all major species for juvenile production and in later stages in coastal farming, mainly for trout and salmon. It is common to use low-temperature water from wells, 20–50°C, and mix it directly with groundwater or seawater (Pálmason, 2005).

The most recent addition to the aquaculture farms in Iceland is the Stolt Sea Farm, located on the Reykjanes Peninsula, next to the power plant. It began operations in 2013 and is currently preparing to produce up to 2000 tons of Senegalese sole, which is exported fresh to markets in Europe. Utilizing a method unique in Iceland, this fish farm uses warm water and electricity from the nearby Reykjanes geothermal power plant. The warm water used is originally cold sea water, which the power plant employs for their tubular power plant condensers. At the outlet from the power plant the water temperature is 35°C, and a quantity is directed to the fish farm while the rest is directed to the sea. In the fish farm, the water is mixed with cold sea water, which is pumped from wells and used in the rearing tanks. The outcome is ideal water for this kind of production, with the exact temperature of 20°C all year round (Morgunblaðið, 2016).

3.6.2.4. Salt production

In 1773, industrial activities were promoted with assistance from the king of Denmark, which at that time ruled over Iceland. Among other things salt production was initiated in Reykjanes, in the Western fjords. In the beginning everything went well but after two decades, the production was terminated as it proved to be economically unviable (Pálmason, 2005). Currently there are two salt factories in operation in the Western fjords and their focus is on the production of high-quality gourmet table salt.

Norður Sea Salt, which is located in Reykhólar in the Western fjords, has been in operation since 2013. The geothermal energy is used to make brine in tanks and then used again to heat open-air-pans for drying the salt. This use of geothermal energy allows the company to achieve the perfect moisture composition of every salt flake. The geothermal water is waste water at 70°C, sourced from a nearby seaweed processing plant, which came originally from a geothermal well with a temperature of 115°C.

Saltverk, which is located at Reykjanes, the same original location as in 1773, began operating in 2011. They utilize about 10 L/s of 90–95°C hot water from a geothermal well, which is cooled down to 70°C for the salt production process. The annual production is 70–80 tons of salt (Ragnarsson, 2015).

3.6.3. Swimming and Bathing

Geothermal water has been utilized since the early days for domestic purposes such as washing and bathing. This is evident from place names recorded in the Sagas 700–800 years ago. The warm springs, which fascinate tourists today, have been used for centuries for bathing. The medieval Icelandic writer, Snorri Sturluson, who was assassinated in 1241, was also an influential politician and a wealthy man. He lived at the farm called Reykholt, where he built a geothermal pool and piped in geothermal water from a nearby hot spring. Archaeologists have excavated parts of the piping system. According to the early annals, Snorri Sturluson used his pool mainly for bathing and socializing, much like modern Icelanders.

Nowadays swimming pools are located in almost every village. There are around 165 swimming pools in Iceland, of which 140 are geothermally heated (Ragnarsson, 2015). These pools provide similar function as the warm bathing spots the early days (Figure 3.15). People of all ages and professions frequent them for social events in order to unwind after a long day, or to catch up on gossip with

Figure 3.15   Swimming pool at Hofsós, NW-Iceland (Photo: Brynja Jónsdóttir).

friends. The public pools are also well used for sporting activities. Swimming pools play a role in the education system, since swimming is a mandatory subject in all elementary and high schools in Iceland.

Most of the swimming pools are located outdoors and are open all year round, from early in the morning to late in the evening. The swimming facilities are usually planned around several pools, including a larger pool for swimming, with water temperatures around 26–28°C, and smaller hot tubs of 37–45°C for relaxing. Some swimming facilities are also equipped with saunas and steam baths.

A geothermal beach has been created in Nauthólsvík in the capital Reykjavík, where the sea water is heated with the excess from hot water storage tanks.

Geothermally heated water has also been utilized in luxury spas. The most famous of these is the Blue Lagoon, located in a lava field on the Reykjanes Peninsula. The lagoon was originally a wastewater pool from the nearby power plant in Svartsengi and has become a major tourist attraction. The wastewater is partly seawater, very rich in minerals such as silica. In the years that followed, people began to bath in the water and apply the silica mud to their skin and realized that it had positive effects on the skin, especially for those suffering from the skin disease psoriasis. The company Blue Lagoon was formally founded in 1992 and has since then worked systematically to create value from the geothermal seawater. It operates together with the bathing facilities, research laboratories, health clinic, and produces skin care products and cosmetic. A similar concept was made in North of Iceland, Mývatn Nature Baths. It is a man-made lagoon where the water supplies run from a nearby well, 130°C, from the power plant Bjarnarflag. The water in the lagoon is cooled down to 36-40°C, the ideal temperature to bath in.

Hot mud bathing is a popular and widely spread alternative health practice. In 1955, the Natural Health Society of Iceland established the Rehabilitation and Health Clinic in the town Hveragerði in South Iceland. At first, they focused on treating people who suffered from rheumatism and today are run as a comprehensive rehabilitation clinic. Although the methods have changed, the geothermal resources available at the clinic, including hot tubs, steam, swimming pool, and a mud bath, continue to be used to treat people.

3.6.4. Industrial Use

Geothermal water has been used for various industrial purposes through the centuries. Washing is probably the oldest one. In the past, women washed clothes outdoors in warm springs. This household task is now undertaken by machine at home and special factories for wool washing and dyeing came later on.

Drying of various products has been the most common industrial use of geothermal in Iceland. Diatomite plant at Mývatn in North Iceland was among the largest industrial users of geothermal energy in the world. The diatomite was mined from the bottom of lake Mývatn, which is known for its special ecosystem and biodiversity supported by natural inflow of geothermal water. The mining was controversial and suspected to be causing a decline in the biosphere of the lake, which was one of the reasons the plant was closed down in 2004, after almost 40 years of operation.

The Sea Minerals Plant at Reykjanes was in operation for a few years but was closed as it proved to be economically unviable. Several other businesses have been attempted, including the drying imported hardwood and re-treading of car tires (Ragnarsson, 2015).

The seaweed plant Thorverk, in West Iceland, established in 1975, is the largest industrial geothermal user today. It produces several thousand tons of clean dry seaweed meal, each year. About 95% of their production is distributed all around the world and used as fodder or fertilizers and as food supplement for humans. The seaweed comes from the nearby bay of Breidafjördur. The weed is chopped and dried indoors where the air is heated to a maximum of 85°C by geothermal water fed through heat exchangers. The geothermal water comes from three boreholes, each 1000 m deep, located close to the factory. It enters the plant at approximately 110°C and heats the drying air of 400,000 m^3/h. The drying time for seaweed is 2–3 h. Under these conditions, all the organic and biologically active substances in the seaweed are preserved intact. The dried material is milled, sieved, packed, and stored. This procedure ensures that all the minerals and organic substances are preserved, prevents surface oxidation, and browning of the meal (Thorverk, 2016).

The carbon dioxide (CO_2) plant Hæðarendi in Grímsnes, South Iceland, has produced liquid CO_2 for commercial purposes since 1986. The geothermal fluid originates in two gas-rich wells in a nearby geothermal field, with intermediate temperature (160°C) and unusually high gas content (1.4% by weight). The gas discharged by the wells is nearly pure carbon dioxide with a hydrogen sulfide concentration of only about 300 ppm. Upon flashing, the fluid from the well would produce large amounts of calcium carbonate scaling. However, scaling in the well is avoided by a 250 long downhole heat exchanger made of two coaxial steel pipes. Cold water is pumped down through the inner pipe and back up the annulus. During this process, the geothermal fluid is cooled to arrest boiling and rapid degassing and the reverse solubility of calcium carbonate is increased sufficiently to prevent scaling. The plant uses approximately 6 L/s of fluid and produces some 3,000 tons CO_2 annually, which is a large share of the Icelandic gas market. The production is used in greenhouses to enrich the atmosphere, for manufacturing carbonated beverages and in other food industries (Ragnarsson, 2015).

Carbon Recycling International (CRI) is an Icelandic-American company located near the geothermal power plant of HS Orka in Svartsengi. It has since 2012 operated a small plant that uses CO_2 emissions of the power plant to produce methanol to blend with gasoline to fuel cars. Hydrogen used in the process is produced locally by electrolysis of water. The current production capacity is 1.7 million liters of methanol per year. Output from the plant is currently used directly as a blend component for standard petrol or as feedstock for biodiesel from esterified vegetable oil or animal fats. CRI and HS Orka have signed an expanded power purchase agreement, which guarantees the availability of sufficient power for CRI to expand the annual fuel production plant up to 5 million liters per year from about 6000 tonnes CO_2 (Ragnarsson, 2015).

ORF Genetics Ltd. is a privately owned biotechnology company established in 2001. They produce recombinant proteins, such as growth factors and cytokines, using its Orfeus[TM] expression system.

This particular protein is useful for biological and medical research. The company manufactures its own skincare products under the name Bioeffect. According to studies, these growth factors stimulate regeneration of the skin and reduce aging. The producing plant is genetically modified barley, which grows in a 2000 m^2 greenhouse. This production is largely automated. The barley plants are grown in nutrient solution, no humus is used and temperature, light, and humidity are precisely controlled. Energy, both electricity and hot water, which is needed comes from the nearby power plant in Svartsengi (Fréttabladid, 2010).

3.6.5. Production of Electricity

Although the production of electricity from geothermal resources was on the agenda from the middle of the 20th century, the progress was slow as it was easier to harness hydropower for electricity in Iceland. (Table 3.2). In 1969, the first power unit was installed in Iceland, a 3.2 MW$_e$ backpressure turbine at the steam field in Bjarnarflag that was already used for the diatomic plant there. This was followed by the 60 MW$_e$ Krafla power plant, the first large-scale geothermal power plant in Iceland. It was built to provide electricity for the north-east part of the country after larger protests against enlargement of an existing local hydropower plant. As mentioned earlier, the Krafla project was badly affected by repeated volcanic eruption in the Krafla volcano over a nine-year period and the plant first came into full effort in 1999 although the first part came online in 1978. The unfortunate volcanic events in Krafla caused mistrust in geothermal power plants that took long time to overcome. However, during this period, two high-temperature geothermal fields, Nesjavellir and Svartsengi (Figure 3.3) were being developed for space heating in Reykjavík and in the municipalities at the Reykjanes peninsula. Production of hot water started at Svartsengi in 1977 and at Nesjavellir in 1990, in both cases followed by co-generation of electricity. In the case of Nesjavellir the production increased from 60 MW$_e$ in 1998 to 120 MW$_e$ in 2005 in four 30 MW$_e$ condensing turbines. In the case of Svartengi the power production was increased stepwise from 1 MWe in 1978

Table 3.2 Geothermal power plants in Iceland and installed capacity 2013 (Ragnarsson, 2015).

Plant name	Plant size (MW)	Year	Unit size (MW)	No of units	Type	Inlet temp. (°C)	Inlet pressure (bar)	Flow rate (t/h)	Estimated production (GWh/yr)
Krafla	60	1978	30	1	DF	172/122	7.2/1.1	400/130	480
		1997	30	1	DF				
Svartsengi	74.4	1977	1	2	SF	159	5	166	
		1981	6	1	SF	155	4.5	166	
		1989–1993	1.2	7	B	103	0.12	131	611
		1999	30	1	SF	163	5.5	275	
		2007	30	1	DS	198	15	288	
Bjarnarflag	3.2	1696	3.2	1	SF	182	9.5	45	26
Nesjavellir		1998	30	2	SF	188	12	432	
		2001	30	1	SF	192	12	198	960
		2005	30	1	SF	192	12	198	
Reykjanes	100	2006	50	2	SF	210	18	324	800
Hellisheiði	303	2006	45	2	SF	178	8.5	600	
		2007	33	1	SF	124	1.05	315	2400
		2008	45	2	SF	178	8.5	600	
		2010	45	2	SF	178	8.5	600	

Note: All installed units are shown but they all are not necessarily yet in use. The abbreviation for the type of unit refer to double flash (DF), single flash (SF), and binary systems (B). The temperature and the pressure refer to the turbine inlet values.

to 76 MWe in 2007. The power production was initially by small backpressure turbines and binary units using the Organic Rankine Cycle (ORC) followed by two 30 MW$_e$ condensing units.

Contrary to the Krafla power plant, the co-generation at Svartsengi and Nesjavellir turned out to be very successful, which renewed the interest for further harnessing geothermal energy for power production. At the same time, further development of hydropower was strongly criticized by environmentalists, which directed the interest of the power industry toward the geothermal power. It was especially the need for large hydropower reservoirs in the highlands that met with opposition. A similar size of a geothermal plant requires much less space and has much less impact on the biosphere.

In the first decade of the 21st century, two large-scale geothermal power plants were built, the 303 MW$_e$ Hellisheiði plant and the 100 MW$_e$ Reykjanes power plant described earlier in this chapter (Figure 3.3). The Hellisheiði power plant is a co-generation plant producing 303 MW$_e$ of electricity and is designed for 400 MW$_{th}$ for space heating. The plant came stepwise online in 2006–2011 with six 45 MW$_e$ high-pressure condensing units and one 33 MW$_e$ low-pressure unit. The plant suffered from several problems in the first years including pressure decline in the reservoir, causing decreasing power production, induced seismicity caused by large-scale injection into active faults, and annoying emission of H$_2$S. These problems have all been solved, the pressure decline by connecting make-up wells to the plant, the induced seismicity by careful injection plans, and the H$_2$S emission by injection and improved emission scheme.

Six geothermal power plants are now in operation (Table 3.2) and the total production in 2013 was 5245 GWh, which accounts for 29% of the total electricity production in the country (Ragnarsson, 2015).

3.6.6. Tourism

Tourism is a relatively new service industry, which has surged in recent years, from around 500 thousand in 2010 to around 1.3 million visitors in 2015 (Óladóttir, 2016). According to the Icelandic Tourist Board, the great majority of visitors to Iceland in both winter and summer, in the years 2013–2014, stated that the landscape and natural environment had influenced their decision to come and experience the country. Geology, geothermal phenomena, and hot springs are among the natural features that attract and interest people visiting Iceland. That means visitors are experiencing energy-related activities during their stay.

The energy companies in Iceland, which utilize geothermal energy, have become aware of the interest among visitors in sustainable green energy and so combine their production and tourism. There are many opportunities for intensive cooperation between those two sectors, tourism and energy. The power plants in South Iceland have established visitor centers or exhibitions within their plant. At Hellisheiði Power Plant, ON Power presents an exhibition, which explores geothermal utilization in Iceland. Likewise, HS Orka, has created a visitor center, which describes our solar systems from its first beginnings. And in the North has Landsvikjun as well opened visitor center during the summer time.

3.6.7. Geothermal Resource Parks

Geothermal resource parks are a relatively new concept in Iceland. It is a name over parks or fields, which encompass areas with energy resources, which are exploited by one or more companies. The objective is to maximize the energy potential for joint use where, for example, the effluent from one company becomes the raw material for another. One such park has been established on the Reykjanes Peninsula in Southwest Iceland, where the power company HS Orka operates two power plants. Another park is in creation by ON Energy, the operator of the Hellisheidi power plant in Southwest of Iceland.

At Svartsengi, the core operation has been the production of electricity and hot water. The first company in the Park to be developed was the Blue Lagoon, not long after the first power plant was commissioned. The slogan of the Park is "society without waste," indicating that the resources are utilized to the fullest extent, in as a responsible manner as possible and for the benefit and further progress of the community. The concept is that there is no such thing as waste, only raw materials that are valuable resources to be used in a wide range of production. As each of the companies of the Park directly utilize two or more resource streams from the geothermal plants, they must, therefore, be located in the vicinity of the power plants. The companies create a complex where science, innovation, and development lead to new products. The resource streams in the park are hot water, sea water, groundwater, electricity, geothermal steam, and CO_2. The main products of the Park are hotel resorts, fish drying, fish farming, cosmetics manufacturing, biotechnology, and aquaculture (Resource Park, 2016).

3.7. The Way Toward the Geothermal Society

In the preceding chapters, we have described the geothermal development in Iceland and how the country benefits form the extensive geothermal utilization. The 90 years of the geothermal development has, however, not always be an easy road to follow. Many types of technical as well as nontechnical issues had to be faced, including problems with scaling and corrosion in the district heating systems, shortage of energy and limited energy delivery security, various technical drilling problems, the risk of unsuccessful drilling, proper management of the production fields, and sustainability of the production. There were two main reasons for the successful development of the geothermal district heating sector, the general public acceptance, and the governmental policy.

The public acceptance is based on several factors that affect the everyday life of people. First, the price level of heating with geothermal energy is much lower in the long run compared to other alternatives for heating like electricity or fossil fuel. Due to the high initial investment cost of the district heating systems and the exploitation cost, the energy price from geothermal can be rather high during the first 10–20 years but then it goes to a price level much lower than can be offered from other energy resources. In addition, the convenience of having hot and cheap geothermal water in your home or your home town opens the possibility for all kinds of side utilization as have been described in earlier chapters. Second, the production of geothermal water is rural in nature, there are limits over how long distances the hot water can be piped. This means that the energy production creates local jobs and give economic advantages to the local municipalities. Third, the environmental benefit of the utilization is highly appreciated by the public.

The governmental policy since the 1970s affected very heavily the development of geothermal utilization for heating. The government allocated considerable financial support for geothermal research and exploration and did set up a risk mitigation fund for geothermal drilling and exploration. In principle, the risk mitigation fund gave loans to cover approximately 60% of the drilling and the exploration cost of the developer, usually the municipalities or local farmers. If the drilling turned out to be unsuccessful, the loan had not to be paid back. It was a prerequisite to get such a loan that the drilling was scientifically justified as having reasonable probability for success. At present, about 90% of the homes in Iceland are geothermally heated but the remaining 10% are heated by electricity and subsidized by the government to the level of the upper limits of geothermal customer prices. In order to facilitate further development of geothermal district heating systems, the government offers in advance payment of up to 12 years of subsidies to cover initial investment cost of a new geothermal district heating service.

The large-scale utilization of high-temperature field for electricity production started about half a century later than the exploitation for heating purposes. It has also been more controversial in Iceland than the low-temperature utilization, mainly because of the environmental issues described earlier. The Icelandic power industry is now striving to solve several technical problems associated with the exploitation of the high-temperature fields and to reduce the footprints of the production on the environment.

The geothermal development in Iceland has almost only be conducted by Icelandic companies, scientist and engineers. It has led to a strong and competent energy companies, research institutions, engineering companies and local support industry. It has created jobs all over the country. It has lowered the energy prices, opened up many opportunities for local companies and employment, and created the strong know-how, which is being exported to other countries. The use of geothermal energy is one of the main pillars of the modern well-being society in Iceland.

Acknowledgments

We would like to express special thanks to our colleagues Björn S. Hardarson for his proofreading and correction in English. As well as to Árni Ragnarsson for his good advice and comments. Furthermore, we thank Albert Þorbergsson, Skúi Víkingsson and Magnús Ólafsson for assistance with maps and tables.

References

Ágústsson, K., Ó.G. Flóvenz, Á. Guðmundsson, and S. Árnadóttir, 2012. Induced seismicity in the Krafla high temperature field. *GRC Transactions*, 36, 975–980.

Ágústsson, K. and Ó.G. Flóvenz, 2005. The Thickness of the Seismogenic Crust in Iceland and its Implications for Geothermal Systems. Proceedings World Geothermal Congress 2005, Antalya, Turkey, April 24–29, 2005.

Arason, S, 2001. Nýting jarðvarma í fiskiðnaði. Orkuing 2001. Reykjavík. 135–146.

Ármannsson, H., J. Benjamínsson, and A.W.A. Jeffrey, 1989. Gas changes in the Krafla geothermal system, Iceland. *Chemical Geology*, 76(3–4), 175–196.

Ármannsson, H., Th. Fridriksson, and B.R. Kristjánsson, 2005. CO_2 emissions from geothermal power plants and natural geothermal activity in Iceland. *Geothermics*, 34, 286–296.

Axelsson, G., A. Gudmundsson, B. Steingrímsson, G. Pálmason, H. Ármannsson, H. Tulinus, Ó.G. Flóvenz, S. Björnsson, and V. Stefánsson, 2001. Sustainable production of geothermal energy: suggested definition. *IGA News, Quarterly No. 43*, January–March, pp. 1–2.

Axelsson, G, 2010. Sustainable geothermal utilization- case histories; definitions; research issues and modeling, *Geothermics*, 39, 4, Special Issue on Sustainable Utilization of Geothermal Energy. 2010, 283–291.

Axelsson, G., Th.Jónasson, M. Ólafsson, Th.Egilson, and Á. Ragnarsson, 2010. Successful utilization of low-temperature geothermal resources in Iceland for district heating for 80 years. Proceedings World Geothermal Congress, Bali, Indonesia, April 25–29, 2010, 9 p.

Bessason, B., E.H. Ólafsson, G. Gunnarsson, Ó.G. Flóvenz, S.S. Jakobsdóttir, S. Björnsson, and Þ. Árnadóttir, 2012. *Verklag vegna örvaðrar skjálftavirkni í jarðhitakerfum*. Orkuveita Reykjavíkur, report 2012, 108 p.

Bjarnason, I.Th., W. Menke, Ó.G. Flóvenz, and D. Caress, 1993. Tomographic image of the mid-Atlantic plate boundary in southwestern Iceland. *J. Geophys. Res.*, 98, 6607–6622.

Bödvarsson, G., 1982. Glaciation and geothermal processes in Iceland. *Jökull*, 32, 21–28.

Cohen, K.M., S.C. Finney, P.L. Gibbard, and J.-X. Fan, 2013. The ICS International Chronostratigraphic Chart. *Episodes*, 36, 199-204.

Einarsson, K., K.E. Sveinsson, K. Ingason, V. Kristjánsson, and S. Hólmgeirsson, 2015. Discharge Testing of Magma Well IDDP-1. Proceedings World Geothermal Congress 2015 Melbourne, Australia, April 19–25, 2015.

Einarsson, P., 1978. S-wave shadows in the Krafla caldera in NE-Iceland, Evidence for a magma chamber in the crust. *Bull.Volcanol.*, 41(3), 189–195.

Flóvenz, Ó.G. and K. Sæmundsson, 1993. Heat flow and geothermal processes in Iceland. *Tectonophysics*, 225, 123–138.

Flóvenz, Ó.G., K. Ágústsson, E.Á. Guðnason, and S. Kristjánsdóttir, 2015. Reinjection and induced seismicity in geothermal fields in Iceland. Proceedings World Geothermal Congress 2015 Melbourne, Australia, April 19–25, 2015.

Fréttablaðið, 2010. (2016, January 4). *ORF vinnur að líffærasmíði*. Retrieved from:
http://www.orf.is/media/PDF/07102010_ORF_vinnur_ad_liffaerasmidi__Fbl.pdf.

Fridleifsson, G.O., W.A. Elders, S. Thorhallsson, and A. Albertsson, 2005. The Iceland Deep Drilling Project –A Search for Unconventional (Supercritical) Geothermal Reservoirs. Proceedings World Geothermal Congress 2005, Antalya, Turkey, April 24–29, 2005.

Friðleifsson, G.Ó., Ó. Sigurdsson, D. Þorbjörnsson, R. Karlsdóttir, Þ. Gíslason, A. Albertsson, and W.A. Elders, 2014. Preparation for drilling well IDDP-2 at Reykjanes. *Geothermics*, 49, 119–126.

Fridheimar (2015, December 20). *Horticulture*. Retrieved from: http://fridheimar.is/en/horticulture.

Hjartarson, Á., 2015. Hallmundarkviða, áhrif eldgoss á mannlíf og byggð í Borgarfirði. *Náttúrufræðingurinn*, 85, 60–67.

Ingason, K., V. Kristjánsson, and K. Einarsson, 2014. Design and development of the discharge system of IDDP-1. *Geothermics*, 49, 58–65.

Jakobsdóttir, S.S., 2008. Seismicity in Iceland: 1994–2007. *Jökull*, 58, 75–100.

Kaban, M.K., Ó.G. Flóvenz, and G. Pálmason, 2002. Nature of the crust–mantle transition zone and the thermal state of the upper mantle beneath Iceland from gravity modelling. *Geophys. J. Int.*, 149, 281–299.

Ketilsson, J., H. Björnsson, S. Halldórsdóttir, and G. Axelsson, 2009. *Mat á vinnslugetu háhitasvæða*. Orkustofnun, report 2009/09, 16p.

Morgunblaðið (2016, January 3). *Rækta senegal-flúru á Reykjanesskaga*. Retrieved from: http://www.mbl.is/vidskipti/frettir/2014/11/06/raekta_senegalfluru_a_reykjanesskaga/.

Mortensen, A. Á. Guðmundsson, B. Steingrímsson, F. Sigmundsson, G. Axelsson, H. Ármannsson, H. Björnsson, K. Ágústsson, K. Sæmundsson, M. Ólafsson,

R. Karlsdóttir, S. Halldórsdóttir, and T. Hauksson, 2009. *Jarðhitakerfið í Kröflu. Samantekt rannsókna á jarðhitakerfinu og endurskoðað hugmyndalíkan.* Report in Icelandic with English abstract, Iceland GeoSurvey, ÍSOR 2009/057, LV-2009/111, 206 p.

Óladóttir, O. Þ., 2016. *Tourism in Iceland in Figures. May 2016.* Report. Icelandic Tourist Board, 2016.

Orkustofnun, 2016. *Orkutölur 2015.* Brochure. Orkustofnun, Reykjavík, 2016.

Pálmason, G., 1973. Kinematics and heat flow in a volcanic rift zone, with application to Iceland, *Geophys. J. R. Astr. Soc.*, 33, 451–481.

Pálmason, G., 2005. *Jarðhitabókin. Eðli og nýting auðlindar.* Hið íslenska bókmenntafélag. Reykjavík, 298 p.

Pálmason, G., G.V. Johnsen, H. Torfason, K. Sæmundsson, K. Ragnars, G.I. Haraldsson, and G.K. Halldórsson, 1985. *Mat á jarðvarma Ísland* (Geothermal Assessment of Iceland), Orkustofnun, Reykjavík, OS-085/ JHD-076, 136 p. (In Icelandic with English summary).

Pálsson, B., S. Hólmgeirsson, Á. Gudmundsson, H.Á. Bóasson, K. Ingason, and S. Thórhallsson, 2014. Drilling Well IDDP-1. *Geothermics*, 49, 23–30.

Ragnarsson, Á., 2015. Geothermal Development in Iceland 2010–2014. Proceedings World Geothermal Congress 2015 Melbourne, Australia, April 19–25, 2015.

Resource Park (2016, February 5). *Society without waste.* Retrieved from: http://www.resourcepark.is/.

Schiffman, P., R.A. Zierenberg, A.K. Mortensen, G.Ó. Friðleifsson, and W.A. Elders, 2014. High temperature metamorphism in theconductive boundary layer adjacent to a rhyolite intrusionin the Krafla geothermal system. Iceland.*Geothermics*, 49, 42–48.

Sigurdardóttir, H. and T.A.Thorgeirsson (eds.), 2016. *OR Environmental Report 2015.* Orkuveita Reykjavíkur, 91 p.

Sæmundsson, K., 1979. Outline of the geology of Iceland. *Jökull*, 29, 7–28.

Thorverk (2016, January 3). *Saga Þörungaverksmiðjunnar.* Retrieved from: http://www.thorverk.is/islenska/fyrirtaekid/index.php.

Þórðarsson, S., 1998. *Auður úr iðrum jarðar. Saga hitaveitna og jarðhitanýtingar á Íslandi.* Hið íslenzka bókmenntafélag. Reykjavík, 656 p.

Violay, M., B. Gibert, D. Mainprice, B. Evans, P.A. Pezard, and O. Flovenz, 2009. *Brittle Ductile Transition in Experimentally Deformed Basalt Under Oceanic Crust Conditions.* Geophysical Research Abstracts, v. 11, EGU2009-0, EGU General Assembly 2009.

Wolfe, C.J., I.Th. Bjarnason, J.C. van Decar, and S.C. Solomon, 1997. Seismic structure of the Iceland mantle plume, *Nature*, 385, 245–247.

Chapter 4

The Italian Challenge
for Geothermal Energy

Adele Manzella*, Assunta Donato,
Gianluca Gola, Alessandro Santilano
and Eugenio Trumpy

National Research Council,
Institute of Geoscience
and Earth Resources, Pisa
*manzella@igg.cnr.it

Italy is the first country in Europe and the sixth country in the world for power generation capacity from geothermal resources, and has been the first country in the world to produce electricity from geothermal energy. The resources, which have been used for bathing and spa since ancient times, nowadays are mainly used for electricity generation and air conditioning by means of district heating (DH) and geothermal heat pump systems. All of the power plants in operation are located in Tuscany, in the two "historical" areas of Larderello-Travale and Mt. Amiata. In the year 2015, the gross electricity generation reached 5.9 TWh, with an installed capacity of 915.5 MWe (807 MWe efficient capacity). The first world hybrid plant with biomass and geothermal has been installed in Larderello in 2015. Direct uses are widespread, developed well beyond the geothermal areas of Tuscany, where, however, most of the geothermal DH systems are located. The heat delivered by direct geothermal uses is 10,500 TJ generating 1,300 MW_{th}, with about half of the installed capacity being related to space heating (DHs and individual systems).

Geothermal contribution to the Italian electrical energy capacity is 1% and to national energy demand is 2%. Although geothermal energy is abundant in Italy thanks to favorable geological conditions, and many operators in more than 100 exploration permits are researching it, there are many obstacles to geothermal expansion of power production from the current situation. Geothermal heat pump application is rapidly growing, and is the sector foreseen to have the largest increase in the next years. This chapter describes the geothermal development in Italy, the current situation and targets.

4.1. Introduction

Italy has an abundance of geothermal resources. Natural manifestations are very common in Italy, and humans have left many traces of their love for thermal waters. The most famous are the numerous spas that Romans built, to the point of exporting the idea of bathing and hygiene in many other Mediterranean and European areas. The role of spa in the Italian culture is very rooted and many hundreds of natural springs gave rise to spa and therapeutic use of hydrothermal waters, so that the most common direct use of geothermal heat has been balneology until recently, when it was passed by space heating applications. Italy is also the first country in the world to produce electrical power from geothermal energy.

4.2. Geological and Geothermal Background

Italy has plenty of geothermal resources, both for shallow applications taking advantage of heat pump technology, and from medium (>90°C) to high (>150°C) temperature systems at depth accessible by wells (usually within 3–4 km). These latter systems occur preferably along tectonically active regions either for volcanic and intrusive or fault-controlled systems (e.g., Moeck, 2014; Santilano *et al.*, 2015 and references therein). The suitability of the Italian territory for exploiting geothermal energy is strictly related to the geological conditions that favor the occurrence of both thermal anomalies and hydrothermal circulation.

4.2.1. Geology and its Implication for Heat Flow

Italy is a tectonically active country, being located along the African–Eurasian convergent margin. The Cenozoic evolution was marked by the formation of two orogenic belts, the Alps and the Apennines,

and two oceanic basins, the Ligurian-Provençal and the Tyrrhenian Sea (Carminati and Doglioni, 2012, Figure 4.1 from Carminati
et al., 2004). The geology of this region is one of the most complex
in the world (for a review see Roure *et al.*, 1990; Doglioni *et al.*,
1996; Patacca and Scandone, 2007; Cosentino *et al.*, 2010; Boccaletti
et al., 2011). The strong lateral variations of the tectonics and of the

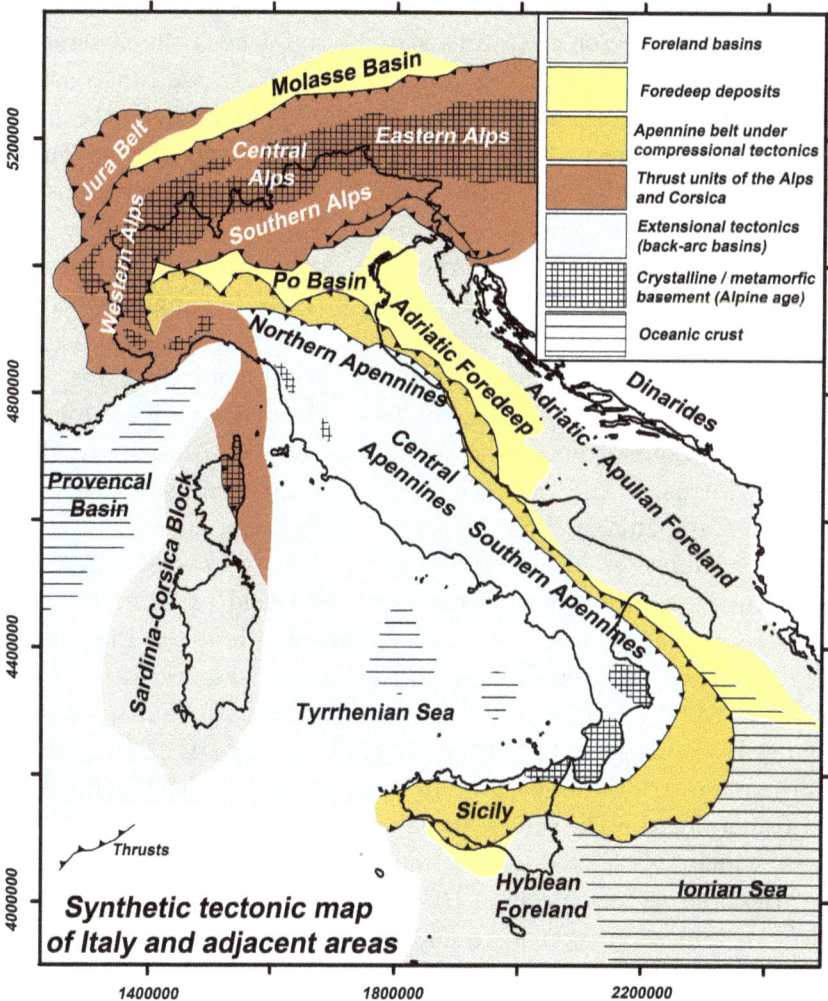

Figure 4.1   Synthetic tectonic map of Italy.

geological features also implied a heterogeneity of the lithospheric thermal regime and of the magmatic activity.

Doglioni (1991) described the Apennine system as a W-dipping subduction with a migration toward East of the back-arc basin, accretionary wedge, and foredeep complex. The subduction hinge retreat caused an eastward migration both of compression at the front of the orogen and extension on the Tyrrhenian side, with the opening of back-arc basins. Leaving aside the processes controlling the kinematics, the related tectonic features strongly influenced the generation of geothermal systems in this area. Schematically, we can consider the central and southern Italy, along the Apennine belt, the most favorable part of the country, while northern Italy in the surrounding area of the Alpine belt is less favorable (with some exceptions for few low-temperature systems).

The inner sector of the Apennine along the peri-Tyrrhenian side shows the most favorable conditions for hosting geothermal resources. This area is characterized by extensional tectonic regime, shallow Moho discontinuity, a reduced lithosphere thickness, and asthenosphere upwelling (Gianelli, 2008). Associated with these features, a widespread magmatic activity developed. Neogene to Quaternary intrusive and volcanic activity occur mainly in Tuscany, Latium, Campania, and Sicilia regions (Peccerillo, 2005; Dini *et al.*, 2005; Carminati *et al.*, 2010), acting as heat sources of very high-temperature systems such as the Larderello field (Tuscany) with a temperature higher than 400°C at shallow depth or the Phlegraean Fields volcanic geothermal system (Campania) (Figure 4.2).

The tectonic complexity of the Italian territory results in a pronounced lateral variability of the surface heat flow (HF), which mimics the main structural features and magmatic domains (Figure 4.3).

Generally, the HF values are low (<50 mW/m^2) in the orogenic belts, which are affected by the time-dependent effect of overthrusting. In these sectors, positive HF anomalies are attributable to the effect of intense erosion in recent times. Rather low values (30–40 mW/m^2) are observed throughout the carbonate-permeable units outcropping in the central-eastern Alpine arc, the central-southern Apennine chain, and the Apulian Platform. Here, the HF is reduced

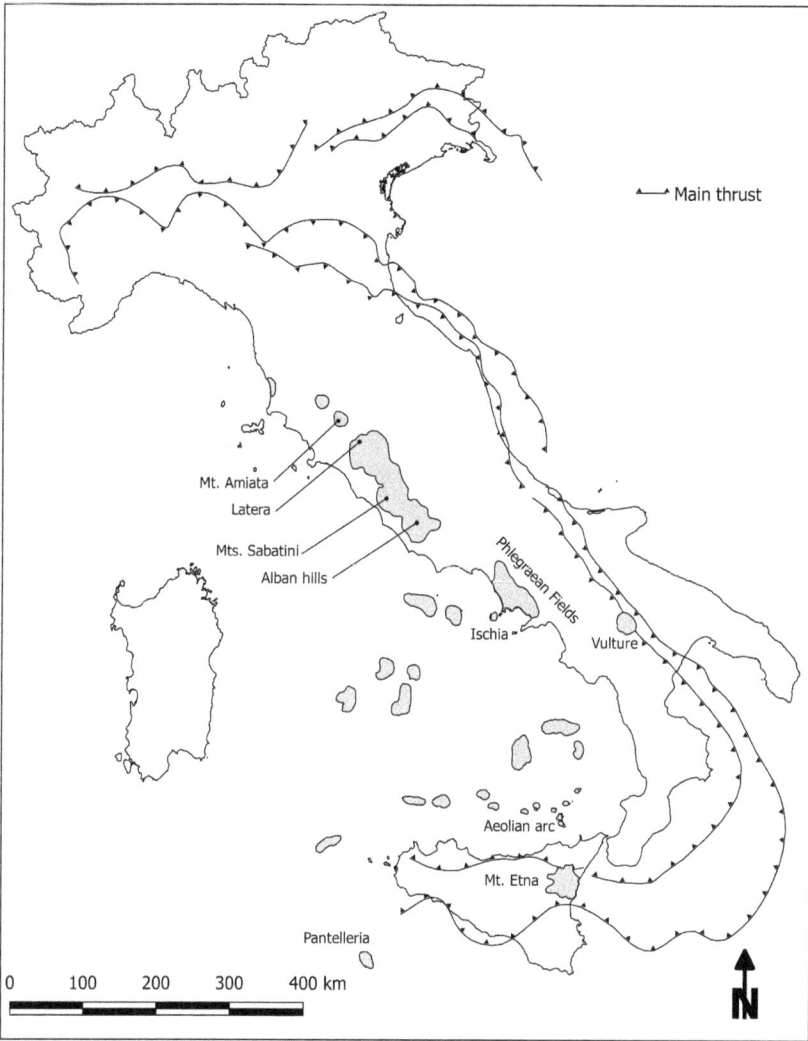

Figure 4.2   Quaternary volcanic outcrop distribution.

by the cooling due to infiltration and circulation of meteoric waters down to large depths. The hydrological circulation is also responsible of the widespread high HF values (80–100 mW/m^2) occurring in the transition zones between the orogenic belts and adjacent basins. The reason of these anomalies can be found in the local rise of warm

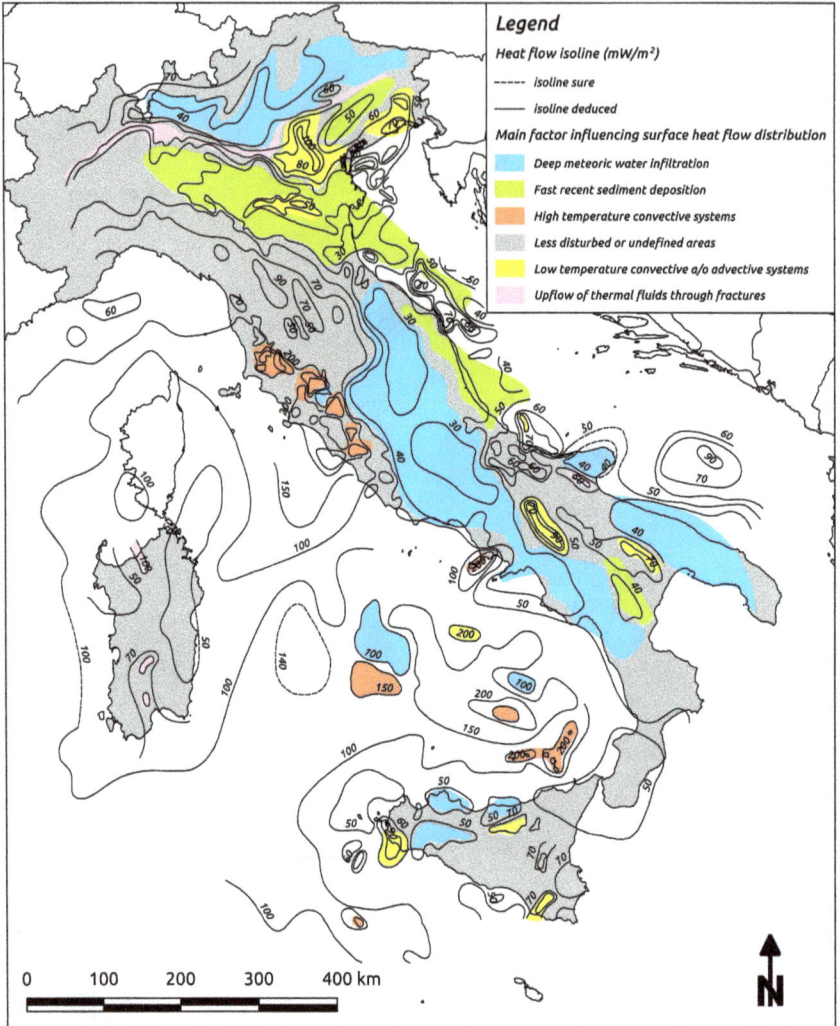

Figure 4.3   Heat flow distribution and its relation to hydrogeological conditions (modified from Della Vedova *et al.*, 2001).

waters from deep formations through faults or lateral discontinuities. Within the basins, in correspondence of structural highs related to the uplift of the carbonate basement, HF increases as consequence of both the thermal conductivity contrast between carbonate rocks and clastic sediments, and the convective heat transfer within the deep-seated permeable units.

The existence at relatively shallow depth of convective hydrothermal systems has been verified in several areas from north to the south of Italian territory and off-shore zones. In the western side of the Apennines, a strong regional HF anomaly (>150 mW/m^2) extends through Tuscan, Latium, and Phlegraean geothermal provinces. The thinning of the lithosphere, accompanied by magmatic intrusion and extrusion processes, are responsible of the high HF values. Similarly, magmatic phenomena related to extension gave rise to the positive anomaly in the Sicily Channel (Pantelleria graben) and Sardinia Island (Campidano graben) as depicted in Figure 4.2. Furthermore, intense extensional tectonics leads to fast sea-floor spreading in Ligurian and Tyrrhenian seas.

To define this complex thermal setting, the temperature measurements and lithological information from thousands of geothermal and oil and gas wells are used as basic parameters and organized in the national geothermal database (Trumpy and Manzella, 2017). Temperature data from geothermal wells are more detailed and reliable, but some thermal data are measured also in oil and gas wells during drilling stops. However, distribution of wells in Italy is not homogeneous, and interpolation of temperature distribution at a depth where wells are not available must always be considered with care. Figure 4.4, for example, depicts the temperature distribution at 2 km depth and location of wells included in the geothermal database (public data).

4.2.2. Geothermal Resources

In Italy, different types of hydrothermal reservoir were recognized in volcanites, sedimentary, and crystalline rocks. The widespread distribution of thermal springs is the surface manifestation of the buried circulation. Hydrothermal circulation was also recognized in deep wells. In most parts of Italy, it is possible to depict a geothermal reservoir at regional scale hosted into Meso-Cenozoic carbonate rocks belonging to the carbonate platforms involved in the

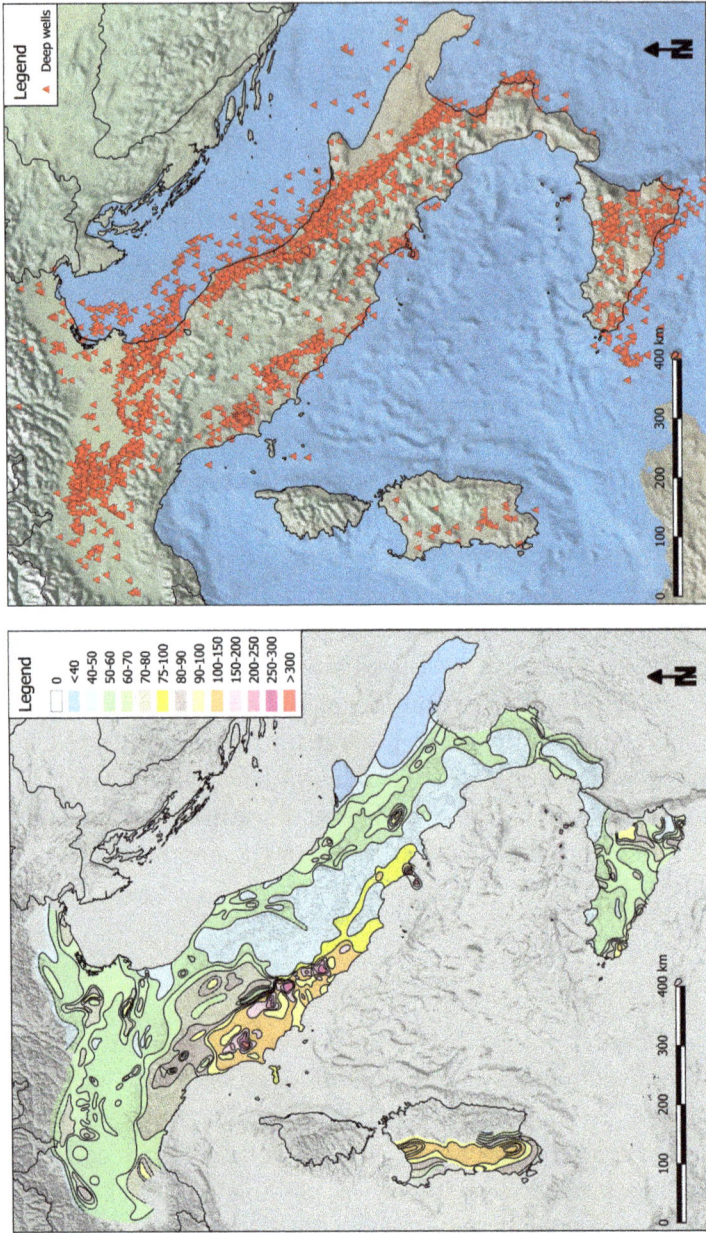

Figure 4.4   Temperature distribution at 2 km depth (left) and location of wells included in the national geothermal database.

Apennine deformation (Montanari *et al.*, 2014; Trumpy *et al.*, 2015). Its temperature and hence its suitability for different technologies (power generation, district heating (DH), and co-generation of heat and power being the most needed) are strongly linked to heat flow.

The last national assessment of deep, hydrothermal resources in Italy was carried out at the end of the 1980s with the completion of the Inventory of the Italian Geothermal Resources. This involved a joint venture including ENEL, ENI, ENEA, and CNR, and later published by Cataldi *et al.* (1995). In the assessment, Italy was divided and ranked in seven categories on the basis of the presence of a regional aquifer of up to 3 km depth and on the temperature range of the fluid. More recently, geothermal potential was computed at the local scale, especially for mining lease requests of power production projects, but at the national scale, the potential has been only roughly estimated. The Italian Geothermal Association (UGI) estimates, on the basis of the temperature distribution, a potential total production from geothermal resources within 5 km depth of 21 exajoule (Buonasorte *et al.*, 2011). Two-thirds of them have temperature below 150°C. In this estimation, resources at temperature suitable for electricity, i.e., those having temperature >80–90°C, can be in many areas, those showing a high surface heat flow and a few more in areas such as Sardinia. In particular with the help of heat pump technology, resources of low temperature can be found almost everywhere in Italy, and can be exploited for direct use of heat.

A comprehensive assessment of geothermal resources in four regions of southern Italy has been recently carried out by the National Research Council. The provided maps show the location of geothermal resources, which can be developed with the current technology. Two maps are related to the development of shallow resources using ground source heat pump (GSHP) systems, providing distribution of areas favorable to the use of open-loop systems (Figure 4.5a) and the realist energy exchange for shallow vertical closed-loop systems (Figure 4.5b, Galgaro *et al.*, 2015). Moreover, the realistic technical potential is computed for deep, hydrothermal resources, on the base of geological conditions (heat in place) and

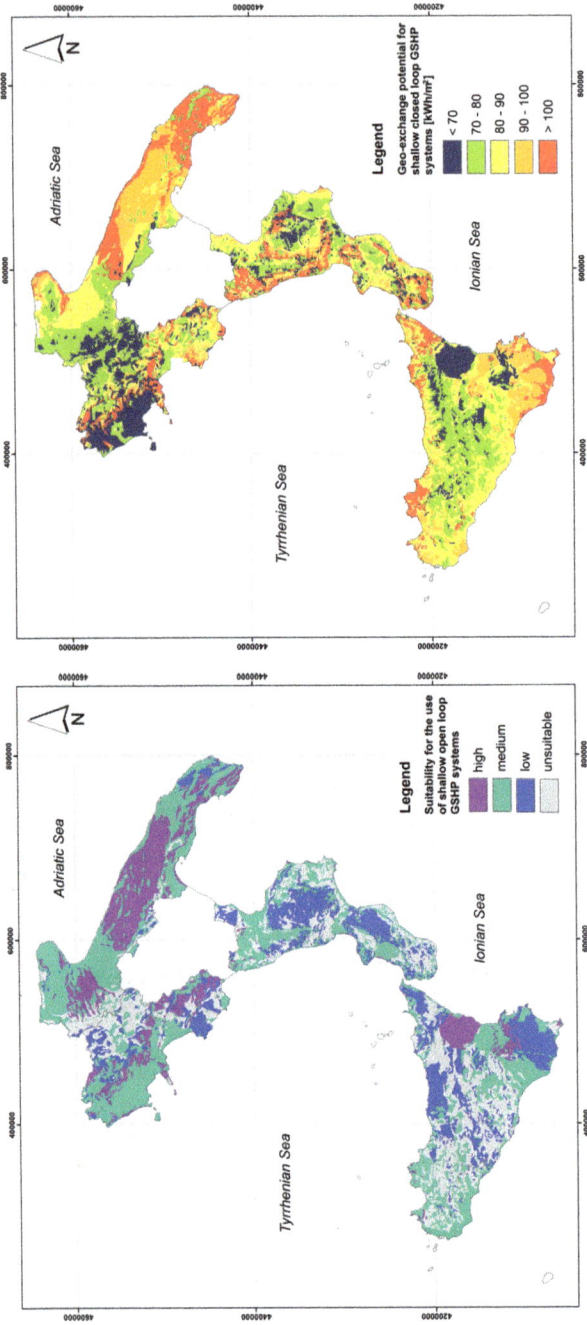

Figure 4.5   Suitability for the use of shallow open-loop GSHP systems (left) and geoexchange energy for shallow closed-loop GSHP systems (in kW/m², right) in four regions of Italy.

the technical requirements for different applications (Trumpy *et al.*, 2016). An example is given in Figure 4.6.

Regarding unconventional geothermal systems, the only assessment available in Italy at the moment is the one provided in the frame of the international project named GEOELEC, and cited in Chapter 1. The map shows that potentiality for enhanced geothermal systems (EGS) power production is large, but at the moment Italy is not interested in developing these resources, mainly due to environmental and social concerns, as discussed in Section 3.6.

4.3. Direct Heat Utilization

In 2015, geothermal direct use applications have reached a total of installed capacity of 1372 MWt, with an estimated heat utilization of 10,500 TJ/yr (Conti *et al.*, 2016). This amount is mostly made by space heating and cooling (H&C), i.e., district H&C systems (DH), and individual systems, in terms of both installed capacity (716 MW_{th}, 52%) and energy use (4445 TJ/yr, 42%), followed by thermal balneology (435 MWt, 3346 TJ/yr), agricultural uses (83 MWt, 683 TJ/yr), fish farming (120 MWt, 1869 TJ/yr), industrial applications (18 MWt, 156 TJ/yr) (Figure 4.7 and Table 4.1). These figures are, however, mostly estimates, especially those at low and very low temperature, since only some geothermal operators are required by law to monitor and declare quantitative figures on direct uses. Thanks to the efforts of the Italian geothermal association (Unione Geotermica Italian [UGI]), recently a geothermal database for direct uses has been established in collaboration with the Italian authorities, and more accurate estimates will be possible in the future years.

Overall, Italy is increasing its share of heat produced by renewable energies, following the policy mandate expressed in the Directive 2009/28/EC for European Union, which finds its Italian expression in the Law Decree 2011/28. Within this increase, geothermal energy plays an important role, being the third thermal renewable energy source, behind bioenergy and air-source in heat pump systems, and supplying about 2% of the total renewable heat consumption. Thanks to the increasing use of GSHP technology, geothermal direct uses in

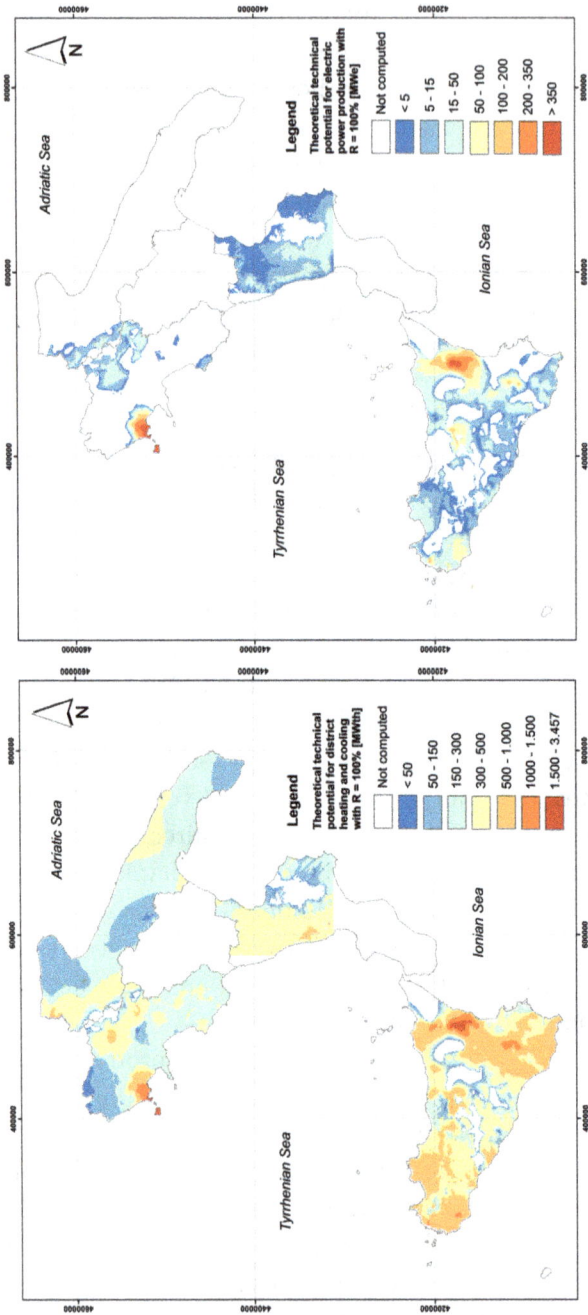

Figure 4.6   Geothermal potential (MWh) for district heating (DH) and cooling systems (left) and power production (right) in four regions of Italy.

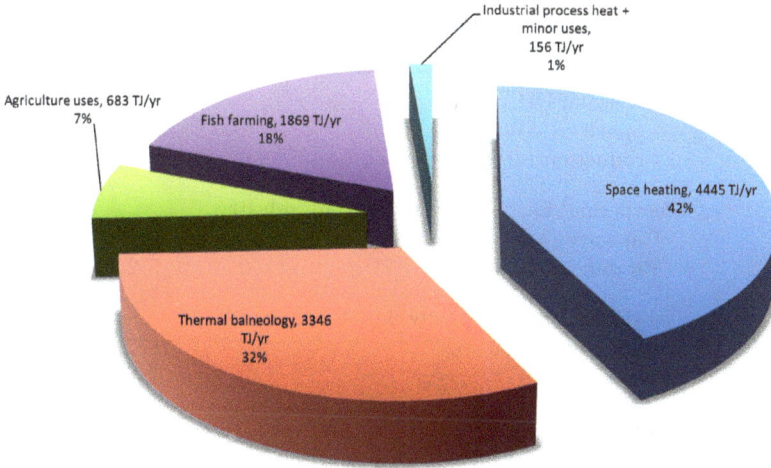

Figure 4.7 Geothermal direct uses distribution in Italy in 2014 by energy use (Conti *et al.*, 2016).

Italy are experiencing an average annual growth rate of over 7%, both in terms of installed capacity and geo-heat utilization. In the total increase of geothermal direct uses between 2010 and 2015 (from 1,000 MW_{th} to more than 1,300 MW_{th}), GSHP accounts for 83% (about 250 MW_{th}), followed by DH and industrial uses (Conti *et al.*, 2016). The latter have seen a renewed interest in particular in the traditional geothermal areas of Tuscany, with application in food and beer industrial processes in the Larderello area and a new leather industry in the Amiata area (Conti *et al.*, 2016). Thermal balneology (the first sector of utilization until 2010) has slightly reduced its relevance. Thermal balneology was provided by the public health service and very used until the 1980s, but the lack of funds has reduced its public support, and many spa systems started to rely more and more on private market related to wellness. Unfortunately, the economic crisis of the last years has reduced the number of customers to about 5% (Conti *et al.*, 2016).

Regarding DH, geothermal energy is used exploiting both hydrothermal fluids with temperature above 40°C without heat pumps, and fresh or warm fluids making use of GSHP technology. Among the DH systems summarized in Table 4.2, two uses GSHP

Table 4.1   Geothermal direct uses in Italy by the end of 2014 (Conti *et al.*, 2016).

Geothermal DH plants	
Capacity (MW_{th})	138
Production (GWh_{th}/yr)	227
Geothermal heat in agriculture and industry	
Capacity (MW_{th})	221
Production (GWh_{th}/yr)	752
Geothermal heat for individual buildings	
Capacity (MW_{th})	577
Production (GWh_{th}/yr)	1008
Geothermal heat in balneology and other uses	
Capacity (MW_{th})	435
Production (GWh_{th}/yr)	929

Table 4.2   Current situation of geothermal DH systems.

Location	Commission date	Geothermal capacity installed (MW_{th})	Geothermal share in total production (%)
Bagno di Romagna (FC)	1983	1.38/1.6	28/32
Castelnuovo V.C. (PI)	1986	11.63	100
Ferrara	1987	14.00	52*
Larderello (PI)	1988	5.00	100
Vicenza	1990	0.70	12
Monterotondo M.mo (GR)	1994	5.15	100
Montecerboli (PI)	1995	5.00	100
Sasso Pisano (PI)	1996	2.33	100
Lustignano (PI)	1996	1.00	100
Serrazzano (PI)	1996	2.50	100
San Dalmazio (PI)	1999	1.00	100
Pomarance (PI)	2003	37.00	100
S. Ippolito (PI)	2003	0.50	100
Santa Fiora (GR)	2005	15.12	100
Montecastelli Pisano (PI)	2010	2.91	100
Milan	2010	20/30	1.2/3
Monteverdi Marittimo (PI)	2014	5.00	100
Montieri (GR)	2014	6.16	100
Grado (GO)	2016	2.00	100

technology: Bagno di Romagna and Milan. In the next years, two DH systems will be completed in the traditional geothermal area of Tuscany, in Radicondoli and Chiusdino. and two other projects have been planned in the same area, in Belforte and Travale.

4.4. Geothermal Power Generation

Italy was a pioneering country in exploiting the potential of geothermal resources for energy power production. After the successful experiment of Piero Ginori Conti with the generation of electricity from geothermal steam in 1904, the first geothermal power plant in the world was built in Larderello, Tuscany, in 1913. The power production has continuously increased since then, with the only stoppage in the 1940s, when facilities were bombed and destroyed by the Second World War events (Figure 4.8).

4.4.1. Current Status

Today, Italy is the sixth country in the world in terms of power generation capacity from geothermal resources. All of the power plants in operation are located in Tuscany, in the two "historical" areas of

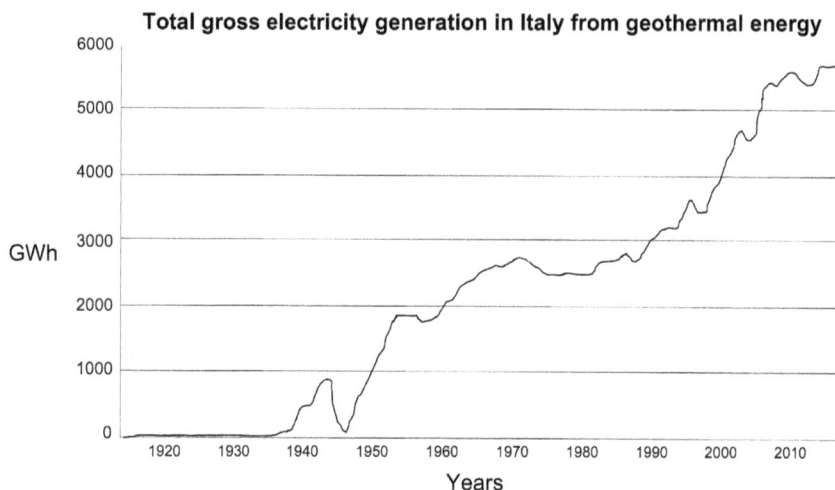

Figure 4.8   Power production in Italy with time.

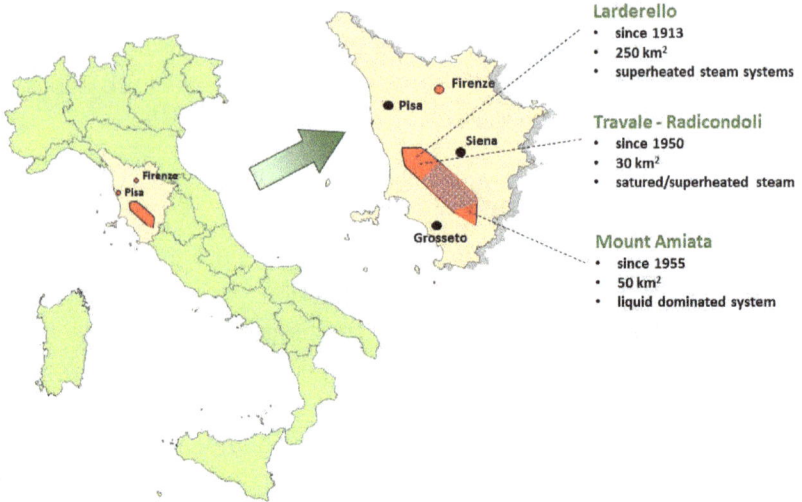

Figure 4.9   Geothermal production areas in Italy.

Larderello-Travale and Mt. Amiata, and are managed by Enel Green Power (Enel GP) (Figure 4.9).

The geodynamic setting and the magmatic activity produce a huge geothermal anomaly in southern Tuscany with maximum peaks centered in the Larderello and Mt. Amiata areas with values up to 1000 mW/m^2 (Baldi *et al.*, 2002). The heat source of Larderello and Mt. Amiata geothermal fields can be ascribed to shallow igneous intrusion belonging to the Tuscan Magmatic Province (TMP) (Santilano *et al.*, 2015 and reference therein). In the geothermal field of Larderello and in the surrounding areas, anatectic volcanic and intrusive rocks are present both in outcrops or in geothermal wells with ages ranging from 4.7 to 1.3 Ma. The intrusive bodies, cored in several deep wells, are granites with ages ranging from 3.8 to 1.3 Ma. Mt. Amiata is a young and extinct volcano (0.3–0.2 Ma). Its volcanic edifice hosts an important cold reservoir, which provides drinkable water to a large part of the surrounding region. Both Larderello and Mt Amiata areas host two reservoirs, the shallow being hosted in sedimentary units and the deep, more extensive reservoir defined by fractures within the metamorphic rocks, at depths of more than

2 km. The deepest well has reached the depth of 4.7 km. The shallow reservoir is steam dominated both in Larderello and Amiata, whereas the thermodynamic condition of the deep reservoir is different for the two areas. At Larderello, superheated steam is present at depths over 3.5 km with temperatures exceeding 350°C and pressure up to 70 bar, whereas the deep reservoir of the Mt Amiata geothermal fields is in a two-phase (liquid and vapor mixture) state with temperatures of 300–350°C and pressure of 200 bar (Conti *et al.*, 2016 and reference therein).

Larderello and Travale/Radicondoli are two adjacent parts of the same deep field, covering a huge areal extension of approximately 400 km² where the deep reservoir has a uniform temperature (350°C) and almost the same pressure (40–70 bar in Larderello, 8–20 bar in Travale-Radicondoli). The field produces superheated steam at a rate of 850 kg/s in Larderello and 300 kg/s in Travale/Radicondoli. The exploited area of Larderello covers about 250 km². Its 200 wells provide fluid to 23 units with 594.5 MW of total installed capacity. On the southeast, the Travale/Radicondoli area covers about 50 km², with 29 wells providing superheated steam to eight units having a total of 200 MW installed capacity.

Mount Amiata area includes two water dominated geothermal fields: Piancastagnaio and Bagnore. Presently there are seven units with 121 MW of installed capacity: four in Bagnore and three in Piancastagnaio. The two-phase fluid that is produced is separated at wellhead at 20 bars. The geothermal fluid is naturally rich with noncondensible gas, which amounts to 6–8% by weight of steam.

Table 4.3 list all plants in operation in the two geothermal areas of Larderello-Travale/Radicondoli and Mt. Amiata (from Conti *et al.*, 2016).

All power plants in operation are dry steam type in Larderello-Travale/Radicondoli. A pioneer binary plant was tested in 2001, but it proved uneconomical: an actually running binary power plant was installed in Mt. Amiata to use the liquid phase after the primary flash of geothermal fluid, and increasing capacity of 1 MWe. The temperature and pressure of the secondary flash that feeds the binary unit are strictly controlled, in order to avoid scaling.

Table 4.3   Power plants in operation in Italy.

Locality	Plant name	Year commissioned	No. of units	Type	Total capacity installed (MWe_e)	Total capacity running (MWe_e)	2014 production (GWh_e/yr)
Larderello	Valle Secolo	1991	2	Dry steam	120	104,8	891
Larderello	Farinello	1995	1	Dry steam	60	53,1	431
Larderello	Nuova Larderello	2005	1	Dry steam	20	15	130
Larderello	Nuova Gabbro	2002	1	Dry steam	20	17,3	150
Larderello	Nuova Castelnuovo	2000	1	Dry steam	14,5	15,1	132
Larderello	Nuova Serrazzano	2002	1	Dry steam	60	42,5	371
Larderello	Nuova Sasso	1996	1	Dry steam	20	8,8	70
Larderello	Sasso 2	2009	1	Dry steam	20	15,1	129
Larderello	Le Prata	1996	1	Dry steam	20	17	144
Larderello	Nuova Monterotondo	2002	1	Dry steam	10	6,1	53
Larderello	Nuova San Martino	2005	1	Dry steam	40	37,5	325
Larderello	Nuova Lago	2002	1	Dry steam	10	10,6	88
Larderello	Nuova Lagoni Rossi	2009	1	Dry steam	20	8,9	75
Larderello	Cornia 2	1994	1	Dry steam	20	11,5	100
Larderello	Nuova Molinetto	2002	1	Dry steam	20	12,3	102
Larderello	Carboli 1	1998	1	Dry steam	20	11,2	97
Larderello	Carboli 2	1997	1	Dry steam	20	13,1	109
Larderello	Selva	1997	1	Dry steam	20	12,9	112
Larderello	Monteverdi 1	1997	1	Dry steam	20	17,2	149
Larderello	Monteverdi 2	1997	1	Dry steam	20	15,4	134
Larderello	Sesta	2002	1	Dry steam	20	9,5	83
Travale-Radicondoli	Nuova Radicondoli	2002	1	Dry steam	40	33,2	275

(Continued)

Table 4.3 (Continued)

Locality	Plant name	Year commissioned	No. of units	Type	Total capacity installed (MWe)	Total capacity running (MWe)	2014 production (GWhe/yr)
Travale-Radicondoli	Nuova Radicondoli GR 2	2010	1	Dry steam	20	18,1	151
Travale-Radicondoli	Pianacce	1987	1	Dry steam	20	17,2	7
Travale-Radicondoli	Rancia	1986	1	Dry steam	20	19,8	172
Travale-Radicondoli	Rancia 2	1988	1	Dry steam	20	18,8	163
Travale-Radicondoli	Travale 3	2000	1	Dry steam	20	14	121
Travale-Radicondoli	Travale 4	2002	1	Dry steam	40	33,1	287
Travale-Radicondoli	Chiusdino 1	2010	1	Dry steam	20	18,4	161
Mt. Amiata	Bagnore 3	1998	1	Single flash	20	19,2	161
Mt. Amiata	Gruppo Binario Bagnore 3	2013	1	Binary (Organic Rankine Cycle [ORC])	1	0,8	6
Mt. Amiata	Bagnore 4	2014	2	Single flash	40	29,7	21
Mt. Amiata	Piancastagnaio 3	1990	1	Dry steam	20	20,3	176
Mt. Amiata	Piancastagnaio 4	1991	1	Dry steam	20	19,5	169
Mt. Amiata	Piancastagnaio 5	1994	1	Dry steam	20	19,5	169
Total			37		915.5	736.5	5914

The two periods of steep increase of power production visible in Figure 4.8 depict the evolution in the development of geothermal reservoirs in Tuscany. The first period, from 1930s to the mid-1970s, is related to the exploitation of the shallow carbonate reservoir, limited to 1 km depth. The second one started in the 1980s onwards, when the fluid production has been increased thanks to the positive results of the deep drilling activity that revealed the presence of a deep reservoir hosted in crystalline (metamorphic and granitic) rocks, and to the artificial recharge of the depleted reservoirs by means of the reinjection of water and condensed steam. Reinjection, which proved to be particularly beneficial in the Larderello and Amiata area since it increased the reservoir pressure and, accordingly, the steam production, and is under testing in the Travale-Radicondoli area (Conti *et al.*, 2016). Despite the prolonged and extensive exploitation history, the continuous increase of Italian geothermal power production is a main example of effective management.

In the year 2014, the gross electricity generation reached 5.9 TWh, with an installed capacity of 915.5 MWe. Considering the real operating conditions of the plants in the different areas (pressure, temperature, noncondensable gas content in the steam), the total running capacity is 807.4 MWe. The world's first geothermal–biomass combined power plant was installed in Larderello in 2015. The biomass, locally produced, feeds a boiler that superheats geothermal steam and increases efficiency and the output power from 12 to 17.2 MWe (Conti *et al.*, 2016).

All of the geothermal power plants are remotely controlled and operated from a remote control station located in Larderello, where 12 persons in turn continuously monitor operations, ready to shut down and restart any unit in case of need. This solution improved operation monitoring, while reducing the operating costs.

Since 1980, in order to increase the productivity of individual wells after drilling and to preserve it during their production life, some stimulation techniques have been developed and are currently being implemented in order to improve the permeability of fractured zones and to reduce or eliminate the formation damage (skin factor) by means of acid stimulation (Scali *et al.*, 2013). With the experience gained during the operation and maintenance of the wells, different causes of well damage (formation or wellbore) have been identified and different techniques aimed at the recovery of the original productivity have been studied and implemented.

Environmental issues are a main concern in Italy. An abatement system named Abbattimento Mercurio e Idrogeno Solforato [AMIS] mercury and hydrogen sulfide abatement) was developed by Enel GP to remove mercury and hydrogen sulfide present in the noncondensable gases of geothermal fluid, and nowadays all power plants are equipped with AMIS systems and high-efficiency demisters. Drilling and power station operation and maintenance are carefully planned and carried out, minimizing environmental impacts.

4.4.2. The New Market

When the Law Decree 22/2010 liberalized the research and exploitation activity of geothermal resources in Italy, and considering the favorable incentives for renewable sources, almost 120 new requests were processed in 2010 and 2011, mostly for new research permits in medium/high enthalpy geothermal resources suitable for power generation, co-generation, and DH, and several new players tried to enter into the market. The rush was seriously reduced by the uncertainty of the market and the serious acceptability problems of local communities, concerned by environmental issues. As a result, at the moment only a few projects completed the surface exploration and were allowed to drill exploratory wells, and the environmental impact assessment (EIA) procedure required for the mining lease is still ongoing, while the administrations still lack clear, comprehensive and shared guidelines.

In the meantime, the electricity market in Italy changed its support for renewable energies. According to the Bill Law issued in July 2012, starting from 2013 new power plants with a capacity exceeding 1 MWe have no longer be granted with "green certificates" but with an "incentive fee" similar to an all-inclusive fee decreased by zonal price of energy to which additional premiums can be added.

In 2015, the average market price of electricity was approximately 4.7 Eurocents/kWh. The value of the net kWh generated from new or recent geothermal power plants awarded with "green certificates" (those Enel GP already operated) was around 13.7 Eurocent/kWh, while with the new "incentive fee" was 9.9 or 8.5 Eurocent/kWh, for units having installed capacity under or above 20 MWe, respectively (Conti *et al.*, 2016).

4.5. Laws and Regulations

The exploration and exploitation of geothermal resources carried out on land, territorial seas, and the continental shelf in Italy are considered to be public interest and utility and are subject to authorization regimes pursuant to Legislative Decree no. 22 of February 11, 2010. The Legislative Decree No. 22 dated February 11, 2010 represents the main regulatory reference for exploration and exploitation of geothermal energy, currently applied in Italy.

Based on the temperature of the geothermal fluids, this decree classifies geothermal resources in: (a) high-enthalpy resources (fluid temperature >150°C); (b) medium-enthalpy resources (fluid temperature between 90 and 150°C); and (c) low-enthalpy resources (fluid temperature <90°C).

The high-enthalpy resources, which can provide an overall power output of at least 20 MW, are considered of national interest, heritage of the state. The medium and low enthalpy resources, which can provide a power output inferior to 20 MW, are declared of local interest. An exploration permit is an exclusive license, issued by the competent authorities, by which all the operations aimed to verify the existence and consistency of geothermal resources are allowed. The exploration permit is issued, also in co-ownership, to the subjects with adequate technical and economic capacity, simultaneously with the approval of the work program and completion of the EIA procedure, if this is applicable. The permit shall require the submission of a bank or insurance guarantee of the applicant, in proportion with the value of environmental recovery works. The area of the permit may cover a maximum of 300 km². The duration is four years and can be extended for further two years. The deadline for the release of exploration permit is 240 days starting from the date of submission of application. The holder of the exploration permit must correspond to €325 per km² to the competent authority. The holder of the exploration permit who has identified geothermal fluids informs the competent authority that, after having attested the national or local interest of discovered resources, publishes it in the Regional Official Gazette (or another advertising tool indicated by the region) and in the Official Bulletin of Hydrocarbons and Geothermal Resources (BUIG). After a successful exploration, the owner may submit the mining lease request within six months from the identification of the resource (national or local interest) to the competent authority. After six months, the mining lease is released after a competition with other operators. The deadline for the release of the mining lease is 220 days. The mining lease of geothermal resources is issued by the competent authority, has a duration of 30 years after approval of the work program, the geothermal project, and after obtaining the positive outcome of the authorities involved and the positive outcome of the EIA procedure. The release of a mining lease is subject to the submission by the applicant of a bank or insurance guarantee. The owner of a mining lease must correspond to €650 per km² to the competent authority and, for power plants producing in excess of 3 MWe, also €0.13 per kWh to the local municipalities, in proportion to the surface of the field, and €0.195 per kWh to the regions involved in proportion to the occupied area. In addition, the owner must pay to municipalities a contribution of first installation equivalent to 4% of the cost of the installations, for environmental and land compensation.

For administrative functions related to the exploration license and mining lease of local and national geothermal resources, including supervisory functions, the competent authorities are the regions or the authorities delegated by them. In case of discovery of geothermal resources in the territorial sea or the continental shelf, the competent authority is the Ministry of Economic Development (MiSE) in cooperation with the Ministry of the Environment and Protection of Land and Sea (MATTM). The ministry must avail itself of the National Mining Office for Hydrocarbons and Geo-resources (UNMIG) for the monitoring of geothermal activities.

Regulations of exploration and exploitation licensing of geothermal resources are specific object of the Presidential Decree No. 395 dated May 27, 1991 and Presidential Decree 485 dated April 18, 1994.

In addition, the Legislative Decree No. 28 of 2011 introduced and regulated the geothermal fluids at medium and high enthalpy, finalized to pilot plants' experimentation with reinjection of wastewaters in the same reservoir, and with zero emissions, with nominal installed capacity not exceeding 5 MW for each power plant. These latter have been introduced in order to promote research and development of new geothermal power plants with a reduced environmental impact. Each applicant cannot be licensed for more than three power plants.

Since presented and accepted with reserve applications in total exceed the maximum limit of 50 MW, a Directorial Notice dated January 31, 2014 established the acquisition of the technical opinion of the Commission for Hydrocarbons and Mineral Resources (CIRM) and stopped further requests.

The competent authority for geothermal permitting of pilot plants is the Ministry of Economic Development (MISE), in cooperation with the Ministry of Environment, which acquires the agreement with the involved regions. The length of the permit for pilot plants is the same as expected for conventional exploration permits with the obligation to finalize installation, commissioning of the plant, and start the testing within the deadlines. If the testing is successful, the owner may submit a mining lease to the competent region and Ministry of Economic Development (MiSE).

Geothermal projects are subordinated to specific environmental regulations. The legislative decree No. 152 dated April 3, 2006 and subsequent modifications and additions represents the main regulatory in the environmental field and request that both exploration permits and mining leases are subject to the EIA screening. The screening is a verification procedure activated in order to evaluate whether proposed plans, programs, or projects may have a significant impact on the environment. If the competent authority considers that the type and level of effects on the environment entailed in a exploration permit is negligible, it can exclude the permit from the EIA procedure.

4.6. Future Perspectives and Conclusions

Italy is a country with large energy needs. In 2014, they reached 323.5 TWh, of which 86.5% were covered with a domestic contribution and 13.5% with imported energy. The 280 TWh of domestic electricity generation in 2014 was produced with fossil fuels (63.0%), hydro (21.5%) and geothermal, biomass, wind, and solar (15.5%). The 5.9 TWh of geothermal power generation represents 2% of Italian electricity generation and covers 30% of the electricity needs in Tuscany, the region in which all power plants in operation are located, giving a substantial contribution to the green energy generation.

In the Mt. Amiata area, after many years in which all activities were stopped due to pending engagement problems from local communities, Enel GP has resumed drilling, replaced old and installed additional units, and completed AMIS systems installation on all power stations. In the Larderello-Travale/Radicondoli area, a deep exploration program including 3D seismic surveys and deep exploratory drilling provided new drilling targets. In the shallow and most depleted areas of the geothermal fields in operation, different strategies for the optimization of resource management including reinjection and chemical stimulation are increasing steam production and reducing natural decline. Many new areas are under exploration and operators are ready to install 50 MW from pilot plant projects and many tens of MW from standard projects. The new, although partial, inventory of national resources has also highlighted favorable areas for geothermal power production.

Serious acceptability problems with local communities, the difficulties of many administrations to carry out all procedures, the little support to renewables, and the consequent long time to recover the large cost of geothermal installations are slowing down and risk to stop many of the 120 projects that proposed the full exploitation of the most interesting geothermal resources in Italy.

A number of initiatives with the intent of achieving a reduction of environmental drawbacks and an increase of acceptability are going on. Enel GP is working on new design solutions to reduce the noise and visual impact of drilling pads, gathering systems and power plants, and further improving abatement systems, while installing available technology for minimizing emissions in all its power plants. Public engagement is increasing and operators organize public initiatives and dissemination. The government is preparing documents for environmental guidance and for providing areal distribution of geothermal resources for power production on the whole Italian territory.

The results of all these efforts will be more evident in the next years. For the moment, the most reasonable target is the installation of additional 80 MW by 2020 in the traditional geothermal areas of Tuscany, as declared by Enel GP (Conti *et al.*, 2016).

Among the unconventional resources, EGS, magmatic and geopressurized resources have not received much attention up to now in Italy, most probably because they would create large concern for the environment: an increasing number of successful stories in other countries is necessary before taking them into consideration. Supercritical resources are, however, being explored in Italy, thanks to new research projects in this field in collaboration with industry and the international community.

Regarding direct uses, the market of GSHP in Italy is in expansion, following the European trend of increasing the share of renewable thermal energy sources. Geothermal DH networks are also notably expanding, with new projects already ongoing in the traditional geothermal areas of Tuscany and the proposed expansion of some actual network (Ferrara, Grado). District and individual cooling is an important option in Italy, especially in the hottest part of the country, and its geothermal contribution will probably increase in the next years.

Industrial applications are increasing and attracting attention, especially for food processes, and also for nutraceutical purposes combining application for greenhouses and industrial processes.

To support the development of the Italian geothermal market, it is required that:

- the know-how is improved, by education and training for operators, installers, manufacturers, since the geothermal market has important peculiarities;
- dissemination is expanded, and promotion is professionally organized;
- procedures for public engagement is successfully tested and become common practice for new projects;
- incentives and guarantees are tailored for the geothermal market, since the risk, capital and maintenance costs are different for geothermal with respect to other renewable energy sources;
- regulations, both for shallow geothermal installations and, in particular, those related to environmental guidance for shallow and deep geothermal, are clearly defined;
- geothermal potential and its distribution are defined on the whole Italian territory in a modern and uniform way, in order to provide an efficient tool for energy planning at regional and national scale;
- research is enlarged and innovation technology improved in order to solve operational issues, to optimize production, and to solve environmental and social concerns.

Geothermal energy is, for Italy, a precious resource, as recently stated at international level by the Italian government. Its reliability as a continuous, but also flexible, sustainable, and renewable energy resource enriches its contribution. Its complete development should become an important target in the energy planning of the country.

References

Baldi, P., Barbier, E., Buonasorte, G., Calore, C., Dialuce, G., Gezzi, R., Martini, A., Squarci, P., Taffi, L. (2002). ITALY — Geothermal thematic map and geothermal areas. In "Atlas of "Geothermal Resources in Europe". *European Commission Publication 1781 In. L-2985*. S. Hurter and R. Haenel Editors. Luxemburg.

Boccaletti, M., Corti, G., and Martelli, L. 2011. Recent and active tectonics of the external zone of the Northern Apennines (Italy). *Int. J. Earth Sci.*. 100, 1331–1348, doi:10.1007/s00531-010-0545-y.

Buonasorte, G., Cataldi, R., Franci, T., Grassi, W., Manzella, A., Meccheri, M., Passaleva, G. (2011). Previsioni di crescita della geotermia in Italia fino al 2030 — Per un Nuovo Manifesto della Geotermia Italiana. *UGI and Ed. Pacini*, Italy. p. 108.

Carminati, E., Doglioni, C., and Scrocca, D. 2004. Alps versus Appennines. *Soc. Geol. It. Spec.*, p. 141–151.

Carminati, E. and Doglioni, C. 2012. Alps vs. Apennines: The paradigm of a tectonically asymmetric Earth. *Earth Sci. Rev.* 112, 67–96, doi:10.1016/j.earscirev.2012.02.004.

Carminati, E., Lustrino, M., Cuffaro, M., and Doglioni, C. 2010. Tectonics, magmatism and geodynamics of Italy: What we know and what we imagine. In: (Eds.) Beltrando, M., Peccerillo, A., Mattei, M., Conticelli, S. and Doglioni, C., The geology of Italy: Tectonics and life along plate margins. *J. Virtual Explorer*, Electronic Edition, ISSN 1441-8142, 36, paper 9, doi:10.3809/jvirtex.2010.00226.

Cataldi, R., Mongelli, F., Squarci, P., Taffi, L., Zito, G., Calore, C. (1995). Geothermal ranking of the Italian territory. *Geothermics*, 24, 115–129.

Conti, P., Cei, M., and Razzano, F. 2016. Geothermal energy use, country update for Italy (2010–2015), *Proceedings of the European Geothermal Congress 2016 (EGC2016)*, Strasbourg, France, September 19–24, 2016.

Cosentino, D., Cipollari, P., Marsili, P., and Scrocca, D. 2010. Geology of the central Apennines: A regional review. In: (Eds.) Beltrando, M., Peccerillo, A., Mattei, M., Conticelli, S. and Doglioni, C., The geology of Italy: Tectonics and life along plate margins. *J. Virtual Explorer*, Electronic Edition, ISSN 1441-8142, volume 36, paper 12, doi:10.3809/jvirtex.2010.00223.

Della Vedova, B., Bellani, S., Pellis, G. and Squarci, P. (2001). Deep temperatures and surface heat flow distribution. G.B. Vai and I.P. Martini (Eds.), Anatomy of an Orogen: the Apennines and Adjacent Mediterranean Basins, 65–76. 2001 Kluwer Academic Publishers.

Dini, A., Gianelli, G., Puxeddu, M., and Ruggieri, G. 2005. Origin and evolution of Pliocene–Pleistocene granites from the Larderello geothermal field (Tuscan Magmatic Province, Italy). *Lithos* 81, 1–31, doi:10.1016/j.lithos.2004.09.002.

Doglioni, C. 1991. A proposal of kinematic modelling for W-dipping subductions — Possible applications to the Tyrrhenian–Apennines system. *Terra Nova* 3, 423–434, doi: 10.1111/j.1365-3121.1991.tb00172.x.

Doglioni, C., Harabaglia P., Martinelli, G., Mongelli, F., and Zito, G. 1996. A geodynamic model of the Southern Apennines accretionary prism, *Terra Nova* 8, 540–547, doi: 10.1111/j.1365-3121.1996.tb00783.x.

Galgaro, A., Di Sipio, E., Teza, G., Destro, E., De Carli, M., Chiesa, S, Zarrella, A., Emmi, G., and Manzella, A. 2015. Empirical modeling of maps of geoexchange potential for shallow geothermal energy at regional scale. *Geothermics*, 57, 173–184.

Gianelli, G. 2008. A comparative analysis of the geothermal fields of Larderello and Mt Amiata, Italy. In: (Ed) Ueckermann, H.I., *Geothermal Energy Research Trends*. Nova Science Publisher, 59–85.

Moeck, I.S. 2014. Catalog of geothermal play types based on geologic controls. *Renew. Sust. Energ. Rev.* 37, 867–882, doi:10.1016/j.rser.2014.05.032, 2014.

Montanari, D., Albanese, C., Catalano, R., Contino, A., Fedi, M., Gola, G., Iorio, M., LaManna, M., Monteleone, S., Trumpy, E., Valenti, V., and Manzella, A. 2014. Contourmap of the top of the regional geothermal reservoir of Sicily (Italy). *J. Maps*, doi: 10.1080/17445647.2014.935503.

Patacca, E. and Scandone, P. 2007. Geology of the Southern Apennines. *Boll. Soc. Geol. Italy* (*Ital. J. Geosci.*) Spec. Issue 7, 75–119.

Peccerillo, A. 2005. *Plio-Quaternry Volcanism in Italy: Petrology, Geochemistry, Geodynamics*. Berlin: Springer, 365 pp.

Roure, F., Howell, D.G., Müller, C., and Moretti, I. 1990. Late Cenozoic subduction complex of Sicily. *J. Struc. Geol.* 12, 259–266.

Santilano, A., Manzella, A., Gianelli, G., Donato, A., Gola, G., Nardini, I., Trumpy, E., and Botteghi, S. 2015. Convective, intrusive geothermal plays: What about tectonics? *Geoth. Energ. Sci.* 3, 51–59, doi:10.5194/gtes-3-51-2015.

Scali, M., Cei, M., Tarquini, S. and Romagnoli, P. (2013). The Larderello — Travale and Amiata Geothermal fields: case histories of engineered geothermal system since early 90's. *Proceedings European Geothermal Conference 2013 (EGC 2013)*, Pisa, Italy, June 3–7, 2013.

Trumpy, E., Donato, A., Gianelli, G., Gola, G., Minissale, A., Montanari, D., Santilano, A., and Manzella, A. 2015. Data integration and favourability maps for exploring geothermal systems in Sicily, southern Italy. *Geothermics*, 56, 1–16. http://dx.doi.org/10.1016/j.geothermics.2015.03.004.

Trumpy, E., Botteghi, S., Caiozzi, F., Donato, A., Gola, G., Montanari, D., Pluymaekers, M.P.D., Santilano, A., van Wees, J.D., and Manzella, A. 2016. Geothermal potential assessment for a low carbon strategy: A new systematic approach applied in southern Italy. *Energy* 103, 167–181.

Trumpy, E., Manzella. A. 2017. Geothopica and the interactive analysis and visualization of the updated Italian National Geothermal Database. *International Journal of Applied Earth Observations and Geoinformation*, 54, 28–37. DOI: 10.1016/j.jag.2016.09.004

Chapter 5

The Turkish
Geothermal Experience

Sakir Simsek

Hacettepe University, Engineering Faculty
Geological (Hydrogeological) Department
Beytepe, Ankara
ssimsek@hacettepe.edu.tr

Geothermal energy is being researched and rapidly developed all over the world as an alternative energy source for energy requirement in countries where the rich geothermal resources exist because of the geological structures. The first geothermal exploration studies in Turkey were initiated by the Mineral Research and Exploration Institute (MTA) in the 1960s and up to the present 230 geothermal fields have been discovered. At these fields, approximately 2000 hot and mineral springs and exploratory-production wells are present and the total proven geothermal capacity is 8,000 MWt. Total calculated potential is 60,000 MWt. While installed capacity of electricity production was at only Kizildere 15 MWe in 2006, it increased by 43 times at the beginning of 2016 and reached about 650 MWe in 14 geothermal fields. Turkey is ranked as the first country in the world in terms of geothermal power plant installed capacity increase during 2010–2015. Utilization of geothermal energy for heating purposes (house, greenhouse, and thermal facilities heating) has reached 2798 MWt. The energy equivalence of geothermal waters used for balneological purposes in Turkey has reached 1005 MWt at 400 spas. At present, there are 20 geothermal district heating (DH) systems in operation at the different centers, mainly Izmir-Balcova, Narlidere, Kirsehir, and Afyon city centers. In geothermal greenhouse heating applications, a significant increase has been provided in the last three years and has reached 3.93 million m^2. The total installed capacity of direct use of geothermal, including liquid carbon dioxide and dry ice

production (annual capacity 160,000 tonnes) and heat pump applications
are 3262.3 MWt. Thermal tourism capacity is very high in the country
and the related investments are increasing rapidly. The Law of Geothermal Resources and Natural Mineral Waters were issued in June 13, 2007.
At the end of issue of law, support of state and tender of the discovered
fields by MTA, research and exploitation studies have increased rapidly.
However, protecting the exploitation of the field is of great importance.
In this chapter, the geothermal development in Turkey and the current
situation and targets will be assessed.

5.1. Introduction

Turkey is one of the richest countries in Europe for geothermal
potential as shown in Figure 5.1. Significant developments have been
achieved in geothermal electricity production and direct uses (district, greenhouse heating, and thermal tourism) throughout the last
50 years. Geothermal law and its regulations have accelerated the
geothermal activities in Turkey. The introduction of the feed-in tariff

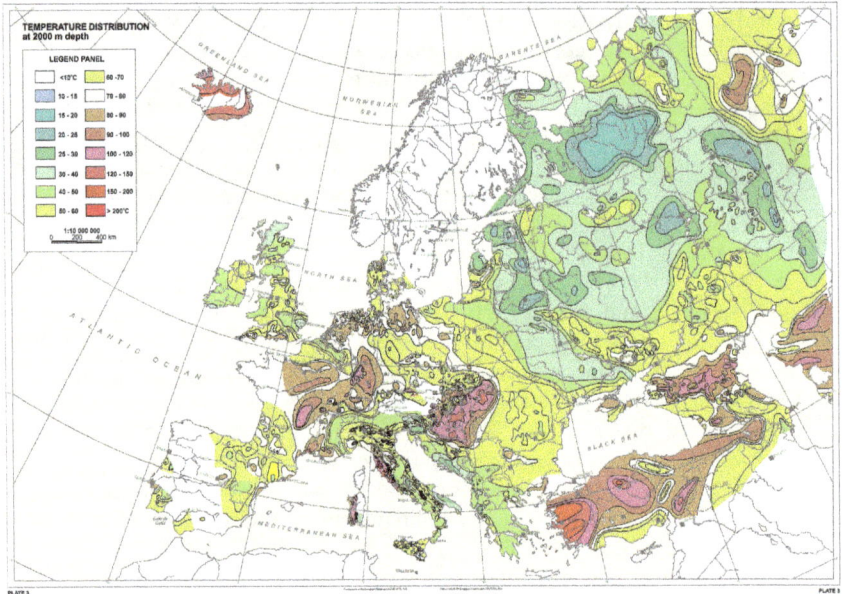

Figure 5.1   Temperature distribution map of 2000 m depth in Europe (Hurtig
et al., 1991).

applications for electricity production, as well as activities of the Mineral Research and Exploration Institute (MTA), have contributed to this acceleration.

Since the 1960s, 230 geothermal fields have been discovered in Turkey. Direct-use applications of geothermal energy have reached 3262.3 MWt geothermal heating, including district heating (DH) (1033 MWt), nearly 3.93 million m^2 greenhouse heating (760 MWt), thermal facilities, hotels, etc. heating 420 MWt, balneological use (1005 MWt), heat pump applications (42.8 MWt), and agricultural drying (1.5 MWt). Geothermal electricity production reached 650 MWe (total 29 geothermal power plants [GPP]) at the beginning of 2016. A liquid carbon dioxide and dry ice production factory is integrated to Kizildere geothermal power plant.

With the existing geothermal wells and spring discharge water, the proven total geothermal heat capacity calculated by MTA is 8,000 MWt (exhaust temperature is assumed to be 35°C).

Most of the development has been achieved in electricity production in the last five years in Turkey, which is the top country for the percentage increase in MWe since WGC2010 (Bertani, 2015). According to the Turkish Ministry of Development's, 10th Development Plan 2014–2018, a total of 750 MWe power production and 4,000 MWt space heating is targeted for the year 2018 (Mertoglu *et al.*, 2015). Thermal facilities for heating and balneological use has gained speed, especially in the last five years (Lund and Boyd, 2015). The released geothermal law contributed to the increase in the geothermal electricity production investments by the Turkish private sector.

5.1.1. Chronological Geothermal Developments

Geothermal developments and first applications in Turkey are given below in chronological order.

Ancient times to the present: Geothermal resources have been used for health, cooking, and some spas have been made that are still in use by Hittites, Romans, Seljuks, Ottomans, and the Turks.

1926: A law related to the quality of hot water for drinking and washing was introduced.

1935: The Mineral Research and Exploration Institute (MTA) was founded.

1947: The first book on geothermal fields was published, entitled (in Turkish) *Turkey Mineral Waters* (Caglar, 1947).

1948: Nuriye Pinar was published on the tectonics and hot and mineral water springs of the Aegean region (Pinar, 1948).

In the 1960s, the increasing importance that the MTA attached to geothermal development meant that geological research and international cooperation projects were founded, while the hot springs and mineral water country inventory was updated.

1962: The first geological survey was carried out in Izmir-Balcova.

1963: The first geothermal well was drilled in Izmir-Balcova.

1966: The first paper was published on thermomineral sources and geothermal energy in Turkey with the contribution of many colleagues and publications by MTA (Erentoz and Ternek, 1966).

1967: Western Anatolia and Denizli geothermal exploration project was started by MTA-UNDP (United Nations Development Programme).

1968: The first high-temperature geothermal reservoirs were discovered at a depth of 540 m with the temperature of 198°C in Denizli-Kizildere.

1974: The first trial geothermal power plant with 0.5 MWe capacity and the first trial greenhouse plant with 1000 m^2 was founded in Denizli-Kizildere by MTA.

1975: The first book by Istanbul University's Faculty of Medicine on Turkey's mineral water is published. National and international Balneology Congresses take place in Turkey.

1979: The first geothermal law draft was prepared by the MTA.

1982: Germencik (232°C) and Tuzla geothermal fields (174°C) were discovered by MTA through drilling.

1982: The Mining Act introduced the first geothermal regulation.

1983: The first geothermal heating system with downhole exchangers was founded in Izmir (Balcova).

1984: The first geothermal power plant with single flash (15 MWe) was put into service by Turkish Electricity Authority (TEK) in Denizli-Kizildere.
1986: The first CO_2 plant was founded in Denizli-Kizildere.
1987: The first geothermal central DH system was started in Balikesir-Gonen city center.
1990: Solutions to hi-level scaling issues were discovered by using chemical inhibitors. Investments increased for heating applications.

In the 1990s, DH applications became widespread and central heating was initiated in Afyon, Kirsehir, Izmir (Narlidere) city centers and provincial centers in Edremit, Bigadic, Salihli, Sandikli, Kizilcahamam, Saraykoy, Simav, Sorgun, Kozakli, and Diyadin.

2000: Following earthquakes on August 17 and November 12, 1999, in Eastern Mamara, geothermal activities were observed in geothermal fields by TUBITAK and other Universities.
2002: The first reinjection activities were applied at the Kizildere geothermal field.
2003: Iller Bank began geothermal research projects and credit support.
2005: The World Geothermal Congress 2005 (WGC2005) was held in Turkey (Antalya).
2006: The first private sector power plant with binary system was established in Aydin (Salavatli-Mege).
2007: The Law of Geothermal Resources and Natural Mineral Waters was issued (Official Gazette numbered 5686 in June 13, 2007) and related regulation entered into force (Official Gazette numbered 26727 in December 11, 2007).
2008: Exploration license applications began after the law was introduced. Exploration and exploitation licenses were granted and present licenses adapted.
2008: Denizli-Kizildere geothermal field was privatized (to Zorlu Inc.).

2009: Exploration and exploitation licenses numbering about 3000 were issued. Exploration and production activities increased.

2009: The first geothermal power plant with double flash (47.4 MWe) was put into service in Aydin-Germencik by Gürmat Inc.

2010–2012: An MTA-discovered geothermal field is put out for international tender to allow for private sector exploration and exploitation.

2013: A 80 MWe power plant is put into service at the Denizli-Kizildere geothermal field.

2014: At the end of 2014, the total installed electricity capacity has risen to 400 MWe.

2015: Total installed capacity has risen to 162.3 MWe at the Aydin-Germencik Omerbeyli field (Gürmat Inc.), and 650 MWe in total in Turkey at the beginning of 2016.

The MTA, banks, various government administrations and municipalities, universities and the private sector continue to provide exploration, exploitation, investment, and financial aspects.

5.2. Geothermal Potential

In Turkey, studies have determined more than 230 geothermal fields, useful at an industrial scale, and about 2000 hot and mineral water resources (spring and well discharge and reservoir temperature), which have temperatures ranging from 20°C to 287°C. Until 2016, a total of nearly 1441 geothermal exploratory, production and reinjection wells have been drilled (Figure 5.2).

These manifestations are located mainly along the major grabens (such as Büyük Menderes, Gediz, Dikili-Bergama, Küçük Menderes, and Edremit Grabens) along the Northern Anatolian Fault Zone, and the Central and Eastern Anatolia volcanic regions (Figure 5.2).

Updated calculations regarding the geothermal heat capacity potential of Turkey is concentrated at 60,000 MWt (Yilmazer, 2009; TJD, 2015). The installed geothermal heat capacity is 3262.3MWt

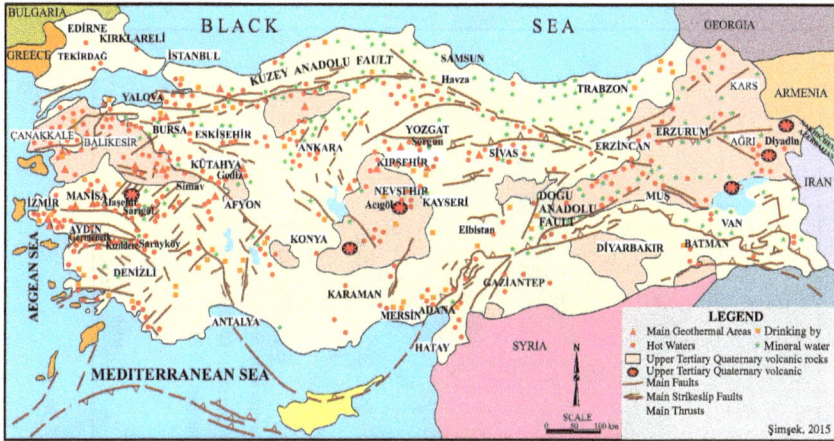

Figure 5.2   Distribution of hot springs, major geothermal fields, and main faults in Turkey.

for direct use (including heat pumps) and 650 MWe for power production. In addition, a liquid carbon dioxide and dry ice production factory is integrated to the Kizildere power plant with a production capacity of 160,000 tons/year according to 2015 data (Figure 5.3).

The total geothermal theoretical electricity potential of Turkey (hydrothermal, 0–3 km) has been calculated as 4,500 MWe (TJD, 2015). According to the estimations and calculations of geothermal electricity, production potential of Turkey could be significant if the buying guarantee of the Turkish Government were 20 years with a feed-in tariff of $20/kWh (TJD, 2015).

5.3. Present Situation of Geothermal Wells

At the beginning of 2016, a total of about 1441 geothermal exploration, production, and injection wells for electricity production and direct use purposes have been drilled in Turkey with a total depth of 1,000,449 m drilled by MTA and the private sector (Dagistan *et al.*, 2015) (Figure 5.4).

In the Büyük Menderes Graben and Gediz Graben geothermal systems, new geothermal fields have been explored (Figure 5.5).

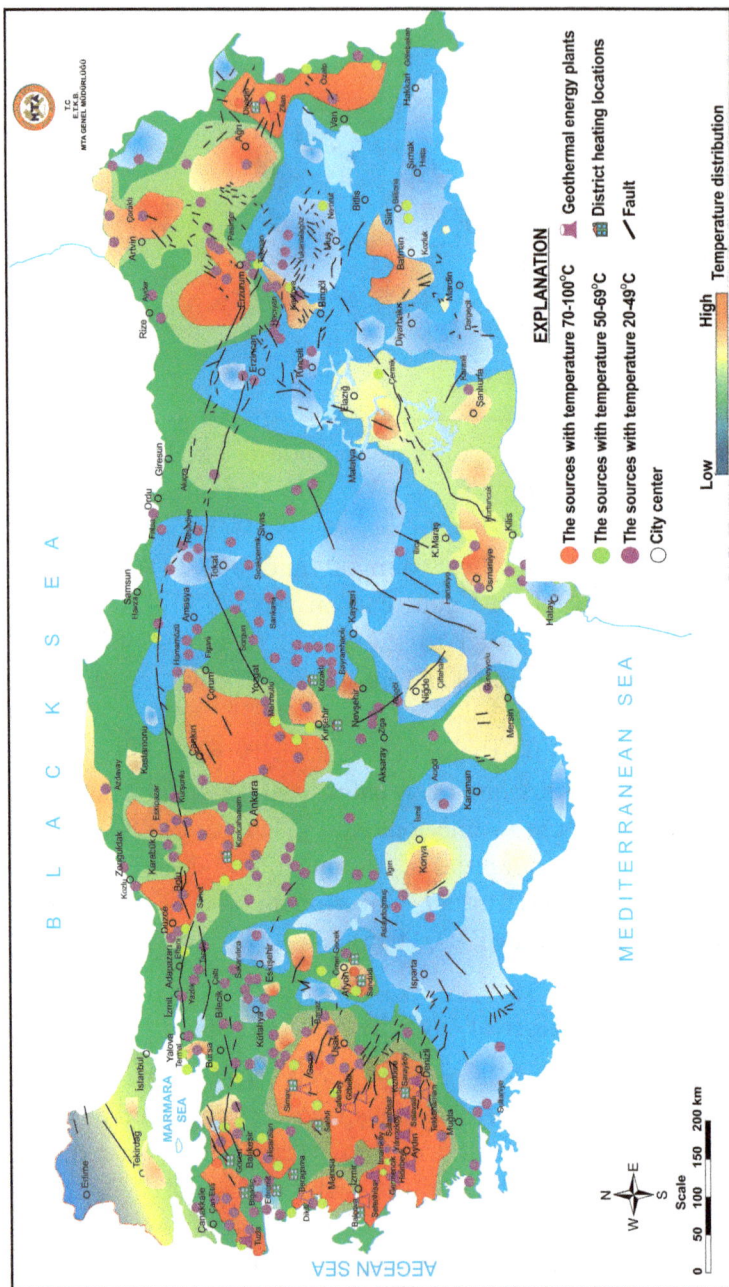

Figure 5.3   Geothermal resources and applications map.

Figure 5.4   Drilling studies at Canakkale–Tuzla geothermal field.

Nearly 80% of the geothermal exploration wells have been drilled in the Western Anatolia region of Turkey.

5.4. High-Temperature Applications

At the beginning of 2016, there exist 29 operating geothermal power plants at 14 geothermal fields in Turkey, which have a total installed capacity of 650 MWe (Table 5.1). Locations of power plants are given in Figures 5.5 and 5.6.

Geothermal Electricity Production Projects from nonartesian geothermal wells have been started at Aydin Gümüskoy, Buharkent, and Denizli-Tekkehamam geothermal fields.

The estimated total projected geothermal electricity use shall be 1065 MWe in the year 2020 (TJD, 2015).

Table 5.1   Utilization of geothermal energy for electric power (2016).

Locality	Power plant name	Year commissioned	No. of units	Type of units	Total installed capacity (MWe)
Denizli*	Kızıldere (Zorlu)	1984/2003	2	1F, 2F, B	15+80
Aydın*	Salavatlı-Dora 1,2,3 (Mege)	2006/2013	3	B	50.86
Aydın*	Germencik (Gürmat)	2009	1	2F	47.4
Canakkale***	Tuzla (Enda)	2010	1	B	7.5
Aydın*	Hıdırbeyli (Maren)	2011/2013	3	B	92
Aydın*	Pamukören (Celikler)	2013	2	B	22.5+22.5
Denizli*	Kızıldere (Bereket)	2007	1	B	6.85
Manisa**	Alasehir Türkerler)	2014	1	B	24
Aydın*	Gümüsköy (BM)	2014	2	B	6.6+6.6
Denizli*	Gerali (Degirmenci)	2014	1	B	2.52
Aydın*	Germencik (Gürmat)	2014	1	B	22.5
Denizli*	Tosunlar (Akça)	2015	1	B	3.5
Aydın*	Pamukören (Celikler)	2015	2	B	22.5+22.5
Aydın*	Germencik (Gürmat)	2015	3	B	22.5+22.5+47.4
Manisa**	Alaşehir (Zorlu)	2015	1	B	45
Aydın*	Umurlu (Kar-Key)	2015	1	B	12
Denizli*	Tekkehamam (Greeneco)	2015	1	B	12.8
Manisa**	Alasehir–Kemaliye (Enerjeo)	2015	1	B	20
Canakkale***	Tuzla (MTN)	2015	1	B	7.2
	Total				**650**

1F = Single Flash, 2F = Double Flash, B = Binary.
* Büyuk Menderes graben; ** Gediz graben; *** Tuzla graben.

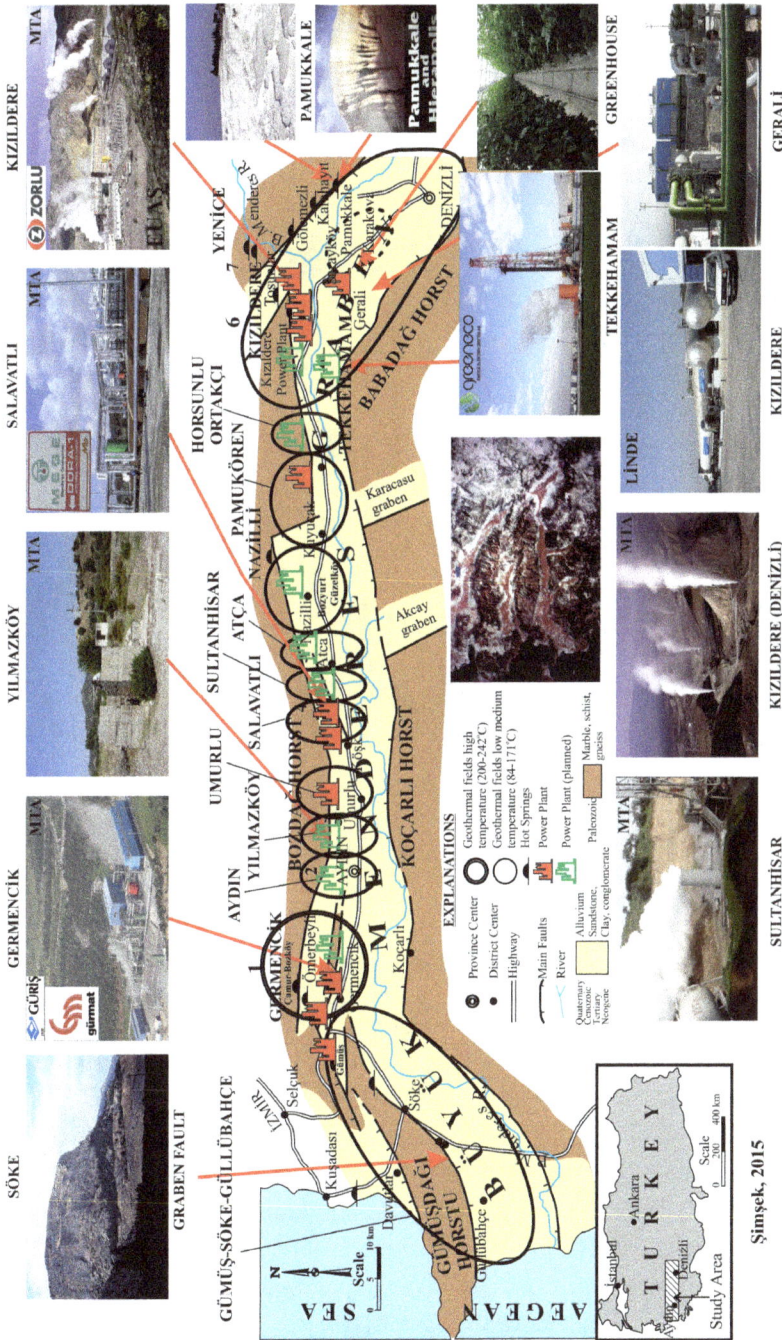

Figure 5.5 Büyük Menderes Graben geothermal systems in the Western Anatolia.

Figure 5.6   Main geothermal fields, heating systems and GPP locations in Western Anatolia.

5.4.1. Kizildere Geothermal Field

The Kizildere geothermal field is located at 40 km west of Denizli City, in the eastern part of the Büyük Menderes Graben (Figure 5.6). It was the first high-temperature geothermal field found in Turkey and was discovered and developed as part of a cooperative project between the Mineral Research and Exploration General Directorate (MTA) of the Turkish Government and the United Nations Development Programme (UNDP). The Turkish Electricity Authority (TEK) installed a power plant with 15 MWe capacity in 1984, which is currently operating (Simsek, 1985; ENEL, 1989).

Research into geology, geophysics (gravity, resistivity, seismic), geochemistry, and gradient drilling were carried out from 1965 to 1968. After that, the first deep well with a depth of 590 m was drilled in order to find the first reservoir. The reservoir temperature is 198°C. To date, 20 deep wells with the depths changing from 370 to 1241 m have been drilled. The field has three reservoirs. The rock type of the first reservoir (170–198°C) is Pliocene limestone and the second reservoir (200–212°C) is Paleozoic marbles-quartzites (Figure 5.7). A third reservoir had been discovered, at depth of 2261 m with a temperature of 242°C in 1998 (Simsek *et al.*, 2005).

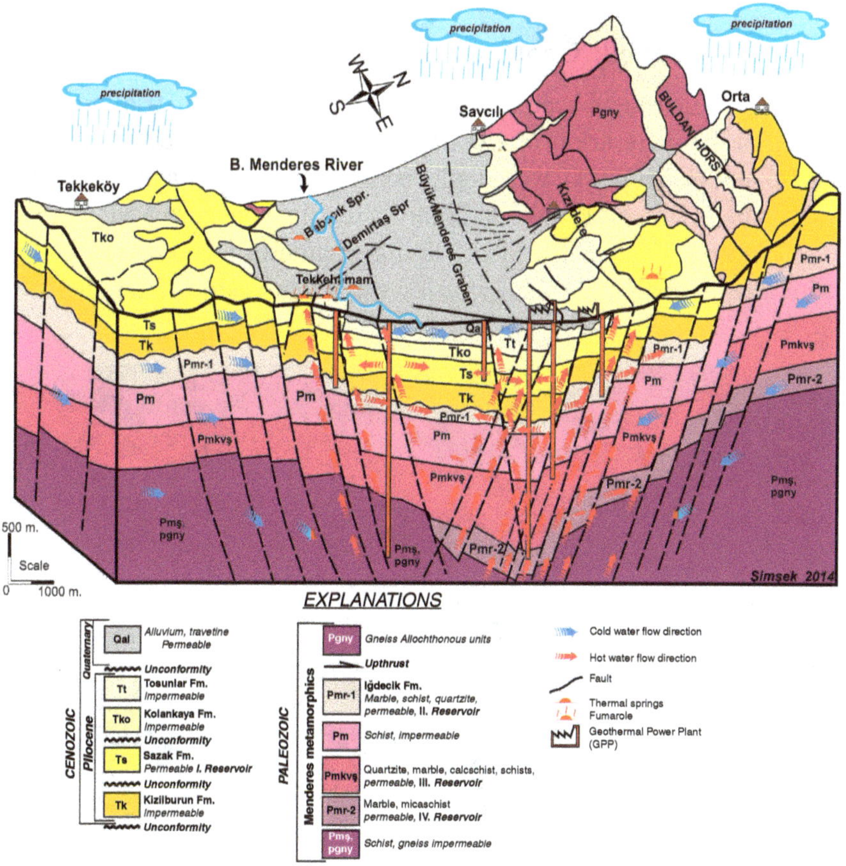

Figure 5.7   Conceptual model of geothermal fields in Büyük Menderes Graben.

The field was acquired by Zorlu Inc. within the framework of the privatization program in 2008, and the power plant has increased to 95 MWe with 80 MWe installed capacity. A fourth reservoir (marble, calcschist) was discovered in the field (240–245°C) and drilling activities are continuing in order to establish new plants.

There is also application for greenhouse heating (100,000 m^2). DH was introduced to Saraykoy town center. In addition, due to noncondensable gas content in the steam, the factory started production of CO_2 in 1985, with a capacity of 120,000 tons/year. The geothermal fluid can also be used in drying and washing of textile products, cooling, touristic and balneological purposes.

5.4.2. Germencik Geothermal Field

The Aydin-Germencik field, which is the second industrial geothermal field for generation of electricity, was discovered by MTA. The first plant (Gürmat-1) was deployed into service as a flash double with capacity of 47.4 MWe in 2009 at the field (Figure 5.8).

This area is placed west of Büyük Menderes Graben and 40 km from the Aegean Sea. The studies started in 1967. After detailed

Figure 5.8　Power plant with a 47.4 MWe capacity in Germencik geothermal field.

geological, geochemical, geomorphological, and geophysical research, the first well with an exploration depth of 1002 m was drilled (Gülay and Gürsoy, 1984; Sahin ve Sener, 1984; Simsek, 1984). The first and second reservoirs have a temperature of 203°C. A total of nine exploration wells with the depth of 285–2398 m were drilled up to 1987. The temperatures of the first and second reservoirs are 203–214°C and 216–232°C (Simsek, 1984). The first and second reservoirs consist of Miocene fractured conglomerates and Paleozoic marble formations respectively (Figure 5.9).

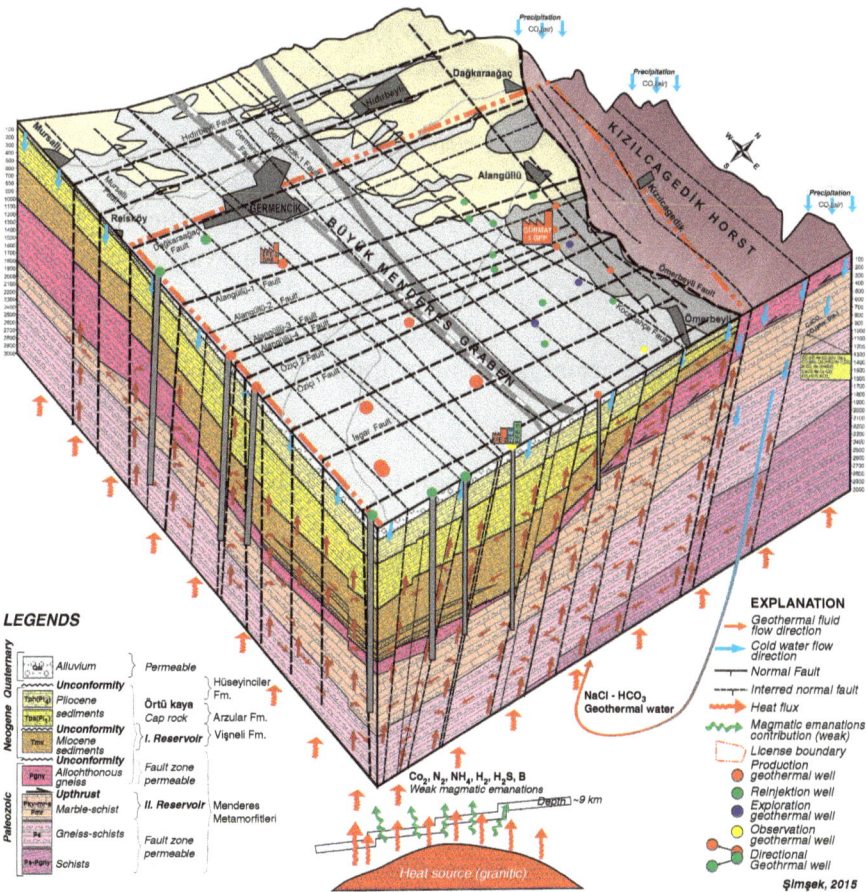

Figure 5.9   Conceptual model of Germencik geothermal field.

The total capacity increased to 162.3 MWe with and three binary power plants (3 × 22.5 MWe) and one double flash plant with capacity of 47.4 MWe which was added by the Efeler group (Efe-1, Efe-2, Efe-3, and Efe-4) in 2014–2015. The deepest and hottest well in Germencik is OB-88 with a depth of 3206 m (276°C).

5.5. Low-Temperature Applications

Turkey is one of the top countries in terms of installed capacity for direct geothermal use. There are 19 geothermal heating systems and 108,930 homes heated by these systems. The operational capacities of the city based existing geothermal DH systems (GHDS) are given in Table 5.2. Today, 40–45°C geothermal waters are used for space heating without heat pumps.

In 2015, geothermal direct use applications have reached a total installed capacity of 3262.3 MWt (Mertoglu and Basarir, 2015). This figure is mostly 1453 MWt (45%) heating (residences, thermal facilities and hotels), 760 MWt (23%) greenhouse heating (nearly 3.93 million m^2), 1005 MWt (31%) balneological use (400 thermal facilities and spas), 1.5 MWt (0,1%) agricultural drying (single application in Kirsehir) and geothermal heat pump applications of 42.8 MWt (1%) (Figure 5.10 and Table 5.2).

5.5.1. Sorgun Field: An Example for Integrated Direct Uses

The Sorgun geothermal field is located in the east of the Sorgun district of Yozgat province. Initially, the hot water requirement for the Sorgun spa was supplied with five shallow wells, which have temperatures ranging from 50°C to 63°C. The first geothermal production well (YS-1) was drilled by MTA in 1988 to increase the amount of hot water for a local spa. Following this, the V-1 and BB-1 wells were drilled in the field. Because of the lignite mining in the Sorgun geothermal field, drilling and galleries were opened so that companies could search for coal mines. Some galleries were flooded with thermal waters due to fault planes cutting through geothermal fluid upwells. For the purpose of thermal water drainage, some wells (YS-3, YS-4

Table 5.2   Current situation of geothermal DH systems.

Location	Residences heated by geothermal	Operation date	Geothermal water temp. (°C)	Investor
Dokuz Eylül Unv. Campus+ Balcova + Narlidere	35000	1983	125–140	Equal partnership of Governorship and Municipality Inc.
Gönen	3400	1987	80	Mainly Municipality Inc.
Simav	12000	1991	137	Municipality
Kirsehir	1900	1994	57	Governorship (Mainly) + Municipality Inc.
Kızılcahamam	2500	1995	70	Mainly Municipality Inc.
Afyon	14000	1996	95	Governorship (Mainly) Municipality Inc.
Kozakli	3000	1996	90	Mainly Municipality Inc.
Sandikli	11000	1998	75	Mainly Municipality Inc.
Diyadin	570	1999	70	Mainly Governorship Inc.
Salihli	8500	2002	94	Municipality
Sarayköy	2500	2002	95	Mainly Municipality Inc.
Edremit	6000	2003	60	Municipality + Private Sector Inc.

(*Continued*)

Table 5.2   (*Continued*)

Location	Residences heated by geothermal	Operation date	Geothermal water temp. (°C)	Investor
Bigadic	1500	2005	96	Municipality
Sorgun	1500	2008	80	Municipality
Yerköy	1500	2009	65	Governorship + Municipality+ Private Sector Inc. Company
Güre	650	2010	55	Municipality
Bergama	450	2010	60	Municipality
Dikili	2000	2010	99	Municipality
Sindirgi	1000	2015	98	Private Sector

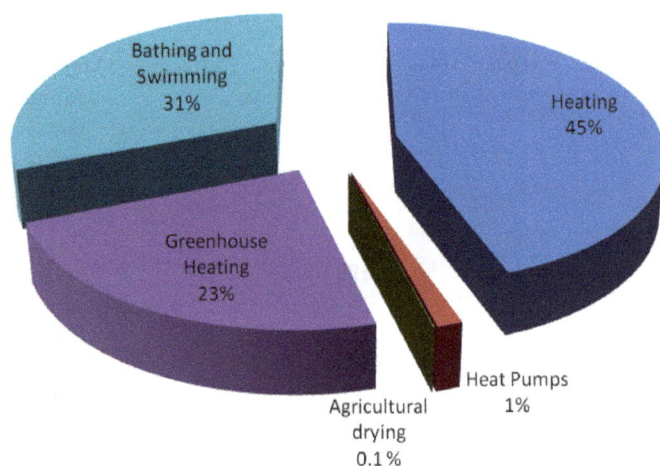

Figure 5.10   Geothermal direct use distribution in Turkey (2015).

and YS-5) were drilled. Following this, five geothermal production and one reinjection well was drilled to determine the potential for hot water and increasing production (Figure 5.11).

Production and reinjection wells with bottom-hole temperatures ranging from 69°C to 85°C were drilled to depths of 104–445 m to obtain thermal water from the field. A reinjection well was drilled

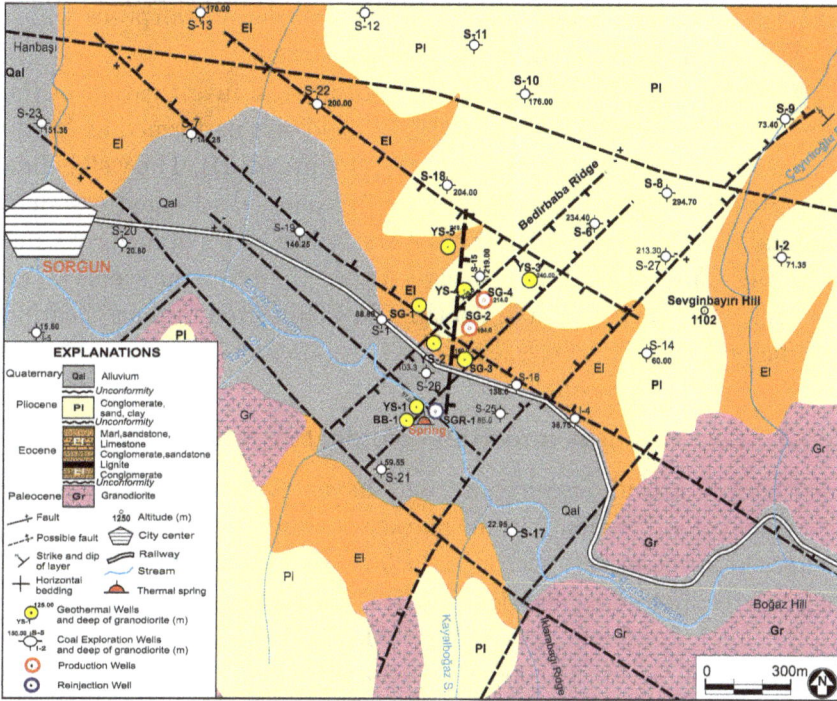

Figure 5.11 Geological map of Sorgun geothermal field and locations of wells.

at the SGR-1 location in the field. Reservoir rocks were hosted by Paleocene granodiorites, whereas Eocene deposits constitute the caprock of the field (Figure 5.11). Thermal waters are of meteoric origin and classified as $NaCl-SO_4$ waters (Simsek *et al.*, 2010). The Sorgun geothermal field is one of the most successful applications for integrated utilizations of thermal waters, specifically central heating, greenhouse heating, thermal tourism and balneology in Turkey. Geothermal wells provided thermal waters for 1,500 residences, a 25,000 m^2 greenhouse and the hot water supply to thermal resorts.

5.5.2. Oil Exploration Wells for Geothermal Possibilities

In southeastern Anatolia, adjacent to the investigated areas, there are many wells drilled by Turkish Petroleum Corporation (TPAO) and Shell Petroleum Companies for oil exploration purposes which do not contain oil but might contain hot waters.

Geological, hydrogeological and geochemical investigation of the Kozluk-Taslidere (Batman) geothermal field was performed in April–May 2001 (Dagistan and Simsek, 2005; (Figure 5.12)). The well Selmo 32A, drilled by the Shell Petroleum Company, discharged taslidere thermal fluid (Figure 5.12). The hot water has 16 l/s of flow rate and a temperature of 83°C. It is observed that the reservoir seems to be between 1700 and 2400 m in depth. The well is used for local heating and balneological purposes.

5.5.3. Greenhouses and Balneological Applications

Geothermal greenhouse applications have reached 3,93 million m^2 and due to good response in the market, greenhouse investments have increased in the private sector in the last few years (Figure 5.13). The major greenhouse applications heated geothermally are as follows: Izmir-Dikili, Bergama: 1,000,000 m^2, Manisa-Salihli, Urganli: 305,000 m^2, Kütahya-Simav: 310,000 m^2, Denizli-Kizildere-Tosunlar: 200,000 m^2, Sanliurfa-Karaali: 474,000 m^2, Izmir-Balcova: 100,000 m^2.

Sixteen million local and 10,000 foreign visitors benefit from balneological utilities in Turkey. Some geothermal fields close to the sea like Izmir-Cesme-Aliaga and Seferihisar, Mugla-Bodrum and Karaada, Aydin-Kusadasi-Davutlar, Balikesir-Edremit, Canakkale-Kestanbol and thermal/sun/history tourism can improve these figures (Figure 5.14).

5.5.4. Heat Pump Applications

Heat pump applications are quite common in the world. The heating of residences and shopping centers in Turkey began with the drawing of ground heat from wells that have approximately 100 m depth. The geothermal heat pump applications including Metro

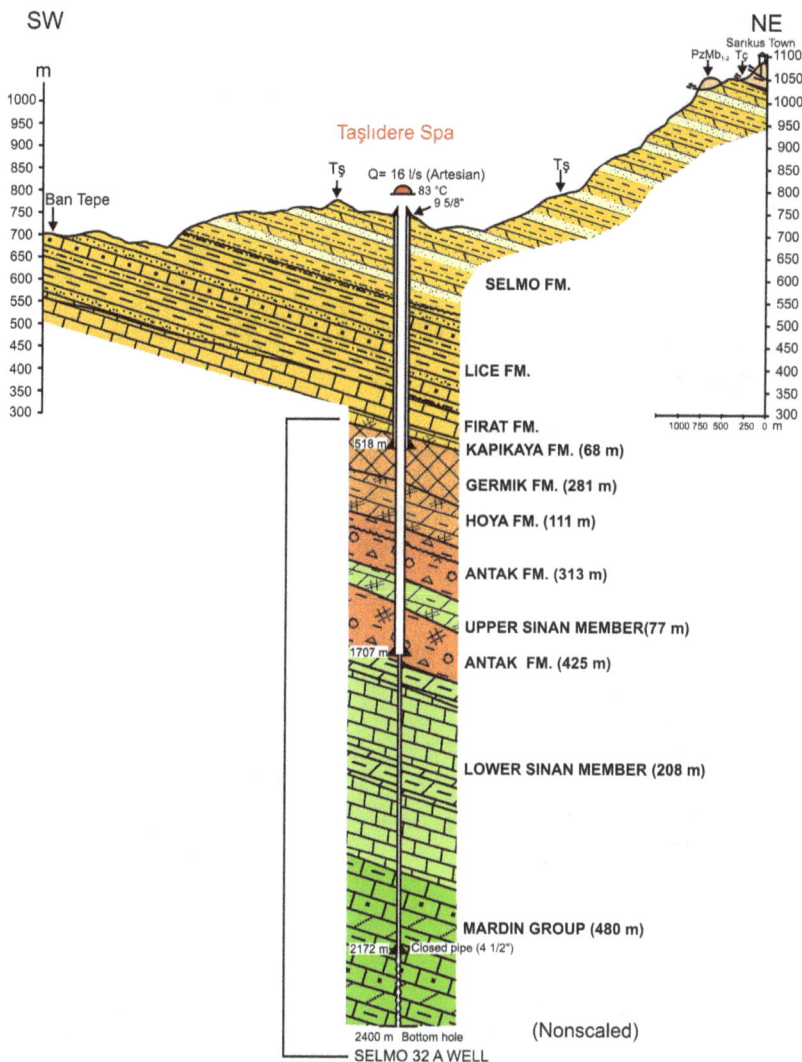

Figure 5.12   Lithological section of Selmo-32A well in Batman; Some of the oil exploration wells using for greenhouse heating purposes will begin at Adiyaman and Carsamba (Samsun).

Meydan M1 Shopping Center in Istanbul (4.6 MWt) and Terme Maris Facility in Dalaman (0.2 MWt), Titanic Hotel in Antalya (8 MWt), Antalya Terracity (12 MWt), Sabiha Gokcen Airport in Istanbul (1.9 MWt) residential heating (1.1 MWt) and others

Figure 5.13   Greenhouse heating in Izmir-Dikili.

Figure 5.14   Balneological applications in Izmir-Cesme geothermal spa and tourism area.

(13.2 MWt, like schools, office buildings, etc.) have gained speed (Figure 5.15). In total, geothermal heat pump applications have reached an installed capacity of about 42.8 MWt (Cetin and Paksoy, 2013). Geological and hydrogeological engineering studies should

Figure 5.15 Heat pump uses in Umraniye Mall (Istanbul).

take place and should report on electricity costs at these facilities in order to increase profitability.

5.6. New Developments and Enhanced Geothermal Systems (EGS) Researches

The anomalies of heat flow density exhibit certain relations to the basic geological and tectonical structure of Turkey. The temperature maps helped to assess the regional temperature field (Tezcan and Turgay, 1991). Very high heat flow up to 150 mW/m^2 is typical for the Paleozoic metamorphic rocks of the Menderes massif in the western part of Turkey. For another example, the surface heat flow distribution of the Marmara Sea region shows values up to 115 mW/m^2 (Pfister *et al.*, 1997). This area is also typical of many boiling and hot water springs, as well as of many other geothermal manifestations, located especially along the graben structures.

Some of the geothermal prospecting areas for which further (EGS and hot dry rocks [HDR]) projects could be carried out are in massifs and young volcanic areas, which are in high heat flow regions. Western and Central Anatolian Massifs are prospect areas. Nevsehir-Acigol field in Central Anatolia is another important prospect area (Carella and Simsek, 2001). The geological and geophysical studies have indicated the presence of an active heat source composed of young extrusives at a shallow depth. Consequently studies directed to HDR can be planned in this region.

The site of a Turkish HDR project at Nevsehir-Acigol Caldera should be selected considering several factors including (Figure 5.12).

- Geological setting of the area;
- Geophysical survey (magnetic, gravity and resistivity map possibilities);
- Convenient topographic conditions;
- Priority for industrial development;
- Limited environmental impact.

Some of the fields are high temperatures measured at depth. One of them is the Alasehir-Kavaklidere geothermal field in Manisa, where 287°C was measured at 2750 m depth in the well. The basement rock at this site is a Paleozoic granodiorite.

The technical and industrial electricity production potential of EGS in Turkey (3–5 km) is calculated is expected to be 25,000 MWe during the next 25-year period and based on a feed-in tariff of \$20/kWh (Mertoglu *et al.*, 2015). In some countries, the feed-in tariff varies between $\sim$€20 and €30 cent/kWh. In Japan, it is in the range of 26–40 yen/kWh (25–\$39/kWh).

5.7. Laws and Regulations

The release of the Law to Use the Renewable Energy Resources for Electricity Production (No: 5346, Date May 10, 2005) has started the acceleration in utilization of renewable energies (geothermal, hydro, wind, biomass and sun). The law gives the price of electricity as incentives for different renewable energy resources. The produced geothermal electricity has got a price of \$10.5/kWh.

Geothermal activities in Turkey are regulated by the Law on Geothermal Resources and Natural Mineral Waters (No: 5686, Date: June 3, 2007) and its Implementation Regulation (No. 26727, Date: December 2007). The geothermal law and its regulations provide solutions to the problems concerning legislative matters and obligations of the exploration and production concession rights, technical responsibility, control and protection of the geothermal areas. The relevant authority is the Ministry of Energy and Natural Resources and the relevant head state entity is the Provincial Special Administration. There are two types of licenses described by law, namely prospecting license and operating license. The former enables its holder to carry out prospecting activities in a specific area based on the project notified to the administration; the latter enables its holder to produce geothermal related water, gas, and steam and use them for energy production, heating, or for industrial purposes.

The geothermal law activated the geothermal activities in all aspects (exploration, drilling, production and utilization) since 2009.

5.8. Environmental Subjects

According to the Law on Geothermal Resources and Natural Mineral Waters and its Implementation Regulation, environmental protection issues have been covered. As required by law,

Figure 5.16   Travertines in Pamukkale is a world heritage site.

reinjection is particularly compulsory for electricity generation and heating projects. Accordingly, reinjection has been made at the field of electricity production and central heating system in Turkey.

In addition, Denizli-Pamukkale, a world heritage site, is protected. Tourism and associated commercial activities have led to physical damage and discoloration of the famous white travertine terraces. To mitigate these environmental impacts, scientific studies were started in 1990s. To protect the terraces and enhance travertine deposition, covered concrete channels have been built to reduce algal growth and a road across the terraces closed (Figure 5.16) (Simsek *et al.*, 2000).

5.9. Conclusion

The 1960s saw the development and update of geothermal project inventory within Turkey. Geothermal electricity applications started in 1984 and geothermal direct use applications started in 1986. In particular, geothermal DH system applications increased rapidly until the year 2000, but geothermal electricity production applications remained single and the same as 15 MWe (Kizildere single flash power plant) until 2007. With the release of the geothermal law and the renewable energy law bringing incentives for electricity production from renewable energy, the Turkish private sector went invested in geothermal power production, meaning geothermal electricity production reached 650 MWe at the beginning of 2016. This can be recognized as big success and expansion. The funds allocated by the Turkish government, exploration activities of MTA, and the tendering of the geothermal fields after one to two exploration wells were drilled in each took an important role in this expansion. WGC2005 organized in Antalya is another important effect of rapid development for geothermal energy in Turkey.

Moreover, thermal tourism (balneology) investments have gained an expansion of nearly 20% in the last one to two years due to the increase in the awareness of the importance of thermal tourism and public health, and to meet the vacation needs of the people during winter time.

Hereafter, Turkey needs to give increased importance to geothermal district cooling. About 70–80% of Turkey's hot geothermal resources are located in Western Anatolia and this area is extremely hot during the summer season and cooling is an important need. To realize cooling geothermally would be important in terms of the environment and would decrease the dependence on fossil fuels and foreign countries.

In Turkey, a second goal could be to give emphasis on EGS to discover its potential and to apply special incentives to EGS investments. According to the estimations and calculations of the Turkish Geothermal Association (TJD-TGA), the geothermal electricity production potential of Turkey could be an significant amount if the buying guarantee of the Turkish government were 20 years with a feed-in tariff of $20/kWh in the coming 25 years.

The main items for exploration and development studies and strategies for the future period are as follows:

- New fields should be excavated, while continued exploration is carried out on present fields to further determine their characteristics and capacities. The required support should be provided to MTA, universities and private sector organizations for their research, development and application projects.
- Because alternative solutions for waste water problems are increased (e.g., reinjection), with regard to the environment geothermal fields can be activated very rapidly.
- Geothermal reservoir protection and management should be realized.
- Deep and high-temperature hydrothermal reservoirs and EGS-HDR should be investigated.
- Scaling and corrosion problems, which affect the management of geothermal energy, have been solved by the injection of chemical inhibitors. As a result, accelerated investment and field activation can take place.
- More geothermal wells should be drilled and the well risk should be supported by the state.

- Determination of utilization possibilities of geothermal fields and planning of these fields in the form of integrated utilization (electricity generation, DH, thermal and balneological applications) and encouragement of the geothermal uses.

The required items for development of Turkish geothermal fields are as follows:

- Support for know-how transfer
- Education and training
- Finance
- Equipment necessities via realization of projects in common with international organizations.

Geothermal energy in Turkey should be used as the main energy source at the regions where it is found, since it is very cheap, clean, and sustainable.

Acknowledgments

The author thanks Turkish Geothermal Association (TJD-TGA) for providing of current data, and Dr. Elif Yilmaz Turali and Mr. Bülent Topuz for preparing of the text.

References

Bertani, R. 2015. Geothermal Power Generation in the World: 2010–2015 Update Report, World Geothermal Congress (WGC2015), No. 01001, Australia.

Carella, R. and Simsek, S. 2001. HDR prospects Italy and Turkey, *International Geothermal Days, Germany 2001*, pp. 301–313, Germany.

Cetin, A. and Paksoy, H. 2013. Shallow geothermal applications in Turkey, European Geothermal Congress, *EGC2013*, Pisa, Italy.

Caglar, K.O. 1947. *Mineral Water and Hot Water Spring in Turkey* (Mineral Research and exploration of Turkey (MTA) Publications, Ankara) (in Turkish).

Dagistan, H. and Simsek, S. 2005. Geological and hydrogeological investigation of Kozluk-Taslidere (Batman) geothermal field, *World Geothermal Congress (WGC2015)*, R-1902, (Eds. R. Horne and E. Okandan) ISBN 975-98332-0-4, Turkey.

Dagistan, H., Kara, I., Peker, B., Celman, O., and Karadaglar, M. 2015. Geothermal Explorations and Investigations by MTA in Turkey, World Geothermal Congress (WGC2015), No. 11094, Australia.

ENEL. 1989. Optimization and development of the Kizildere geothermal field. *Project Final Report.* Pisa, Italy (Unpublished).

Erentoz, C. and Ternek, Z. 1966. *Thermo Resources and Geothermic Energy Studies in Turkey* (Mineral Research and Exploration of Turkey (MTA) Publications, Ankara) (in Turkish).

Gülay, O. and Gürsoy, T. 1984. The Interpretation of gravity data in the Aydin-Germencik geothermal field, *U.N. Seminar on Geothermal Energy*, EP/R.39, Italy.

Hurtig, E., V. Cermak, R. Haenel, V.I. Zui (Eds.) 1991. *Geothermal Atlas of Europe.*

Istanbul University 1975. Turkey Mineral Water (Istanbul University Faculty of Medicine, Chair of Medical Ecology and Hydroclimatology Press. Istanbul) (in Turkish).

Lund, J.W. and Boyd, T. 2015. Direct utilization of geothermal energy 2015 worldwide review, World Geothermal Congress (WGC2015), No: 01000, Australia.

Mertoglu O., Basarir, N., and Saracoglu, B. 2015. Turkey's geothermal potential on EGS-enhanced geothermal system, World Geothermal Congress (WGC2015), No. 31041, Australia.

Mertoglu O., Simsek, S., and Basarir, N. 2015. Geothermal country update report of Turkey (2010–2015), World Geothermal Congress (WGC2015), Proceedings, No. 01046, Australia.

Mertoglu O. and Basarir, N. 2015. Update: geothermal heat in Turkey. Geopower & Heat Summit 2015, Presentations, Istanbul, Turkey.

Pfister, M., Rybach, L., and Simsek, S. 1997. Geothermal reconnaissance of the Marmara Sea region, active tectonics of Northwestern Anatolia, *Vdf Hochschulvertag AG an der ETH Zurich*, ISBN 3-7281-2425-7, pp. 503–537, Switzerland.

Pinar, N. 1948. *Tectonics and Hot and Mineral Water Springs of Aegean Area* (Istanbul University Science Faculty Press, Istanbul) (in Turkish).

Sahin, H. and Sener, C. 1984. Resistivity studies in Aydin-Germencik Geothermal Field. *Utilization of Geothermal Energy for Electric Power Production, Space Heating, United Nations Economic Commission for Europe*, EP/SEM.9/R.38, Florence, Italy.

Simsek, S., Yildirim. N., and Gulgor, A. 2005. Developmental and environmental effects of the Kizildere geothermal power project, *Geothermics*, 34, 234–251.

Simsek, S., (1985). Geothermal model of Denizli, Saraykoy-Buldan area, *Geothermics*, 14, 393–417.

Simsek, S., Günay, G., Elhatip, H. and Ekmekci, M. 2000. Environmental protection of geothermal waters and travertines at Pamukkale, Turkey, *Geothermics*, 29, 557–572.

Simsek, S. and Yildirim, N. 2000. Effects of the 1999 Earthquakes on Geothermal Fields and Manifestations. IGA-International Geothermal Association News No: 40, p. 7–9 Pisa-Italy.

Simsek, S. 1984. Aydin-Germencik-Omerbeyli geothermal field, Utilization of Geothermal Energy for Electric Power Production, Space Heating, United Nations Economic Commission for Europe, pp.1-30, EP/SEM.9/R.37, Florence, Italy.

Simsek, S., Yilmaz E., Koc K., Turker, O., Karakus, H., Bakir, N., and Gulgor, A. 2010. Geothermal Exploration Survey of Sorgun Geothermal Field (Yozgat-Turkey). World Geothermal Congress (WGC2010), No. 1124, Bali, Indonesia.

Tezcan, A.K. and Turgay, M.I. 1991. Heat flow and temperature distributions in Turkey. In: V. Cermak, R. Heanal, and V. Zui (eds.), *Geothermal Atlas of Europe*, Herman Heak Verlag, Gotha, Germany.

TJD. 2015. Geothermal Energy Development Report, Turkish Geothermal Association (TJD), Ankara (Unpublished).

Yilmazer, S. 2009. Determining the potential of geothermal potential of Western Anatolia, 11. Turkey Energy Congress, Izmir (in Turkish).

Chapter 6

Unlocking Geothermal Energy from Mature Oil and Gas Basins: A Success Story from the Netherlands

Jan-Diederik van Wees[*,†,‡], *Maarten Pluymaekers*[*],
Damien Bonté[†], *Serge van Gessel*[*],
and Hans Veldkamp[*]

TNO Energy, Utrecht, the Netherlands
[†]*Utrecht University, Department of Earth Sciences*
Utrecht, the Netherlands
[‡]*jan_diederik.vanwees@tno.nl*

Geothermal development in aquifer settings, targeted beforehand by oil and gas exploration, can benefit largely from putting subsurface information and models in a geothermal context. To this end, we developed a generic resource and performance assessment methodology to identify prospective high-permeable clastic aquifers and to assess the amount of potential geothermal energy. The workflow has been applied for the Netherlands and results are publicly accessible through the web-based portal ThermoGIS (www.thermogis.nl). Public access to the geothermal information system and relevant subsurface data from oil and gas exploration and production (www.nlog.nl) has facilitated spectacular growth of low-enthalpy geothermal projects for greenhouse heating and the number of issued licenses promises an even faster growth in the future.

6.1. Introduction

Geothermal low-enthalpy heat in nonmagmatic areas can be produced by pumping hot water from aquifers at depths typically in excess of 1 km to reach sufficiently high production temperature. Temperature, depth and thickness structure, and permeability of aquifers are key parameters for performance and need to be assessed by geothermal exploration. Geothermal exploration is costly, but can take considerable advantage from a wealth of existing oil and gas data, as oil and gas reservoirs can be potential geothermal aquifers, which is often the case for low-enthalpy geothermal systems, as the critical reservoir properties for hydrocarbon and geothermal exploration partially overlap.

The Netherlands holds a rather unique position in terms of subsurface data availability. Oil and gas exploration has provided us with a legacy of information from over 5000 wells, over 70,000 km of 2D and 3D seismic covering over 60% of the country (Figure 6.1).

All of these data are available in digital form through www.nlog.nl and publicly accessible. The high level of public subsurface knowledge and well-matured exploitation of its hydrocarbon resources provides an important stepping stone for the reliable development geothermal energy.

In 2007, a horticulturist in Bleiswijk was the first to successfully produce hot water from clastic aquifers of around 2000 m depth, which have been explored for oil and gas production. Following this success, interest in geothermal energy spectacularly increased. Geothermal energy in the Netherlands has been growing rapidly over the last decade, with currently over 10 installations developed (Figure 6.2). Issued licenses promises an even faster growth in the future.

The Netherlands setting highlights that rapid geothermal development in mature oil and gas exploration setting can greatly benefit from public availability of data and models of the subsurface. TNO, in its role as geological survey, in the past years facilitated the geothermal development through developing of novel workflows for putting relevant subsurface data and models in a geothermal context for project developers and policymakers. Complementary, public performance screening models have been developed to assess

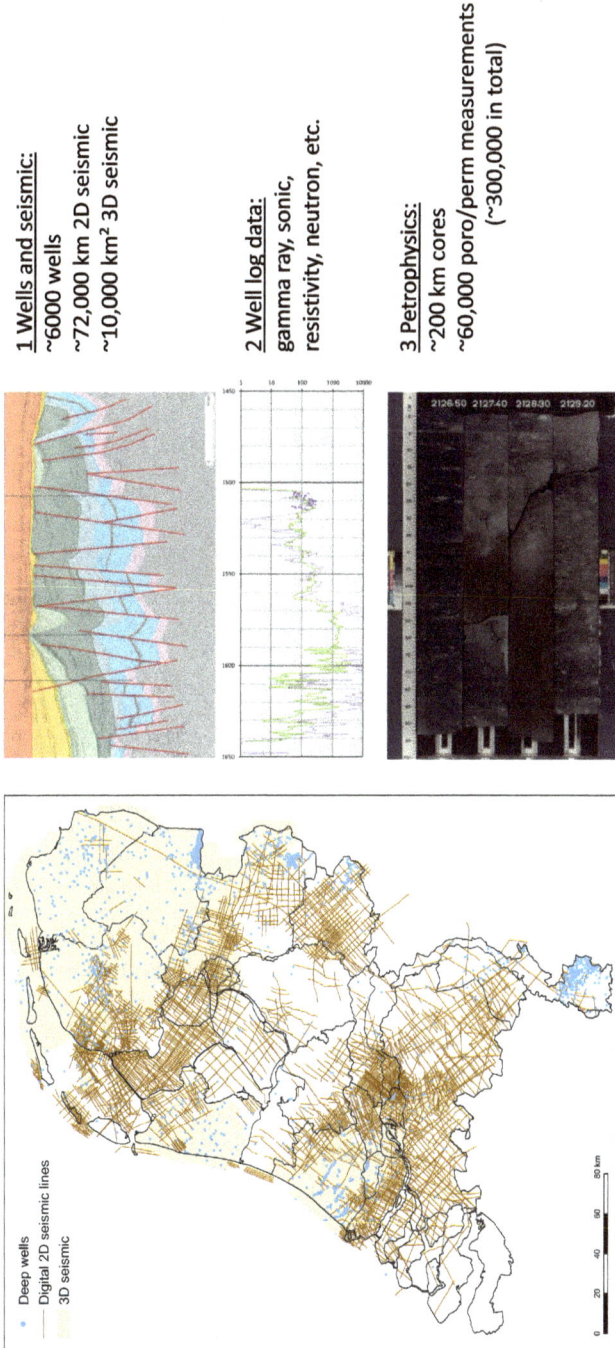

1 Wells and seismic:
~6000 wells
~72,000 km 2D seismic
~10,000 km² 3D seismic

2 Well log data:
gamma ray, sonic,
resistivity, neutron, etc.

3 Petrophysics:
~200 km cores
~60,000 poro/perm measurements
(~300,000 in total)

Figure 6.1 Hydrocarbon exploration data available in the Netherlands, including over 5,000 wells onshore (blue), 2D and 3D seismic lines, log data and core data from wells.

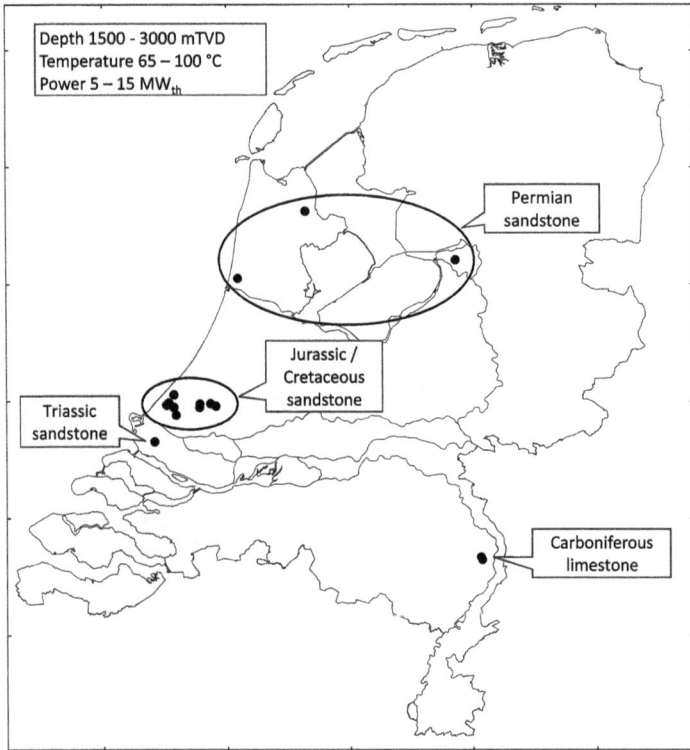

Figure 6.2   Active geothermal plants in the Netherlands (status 2015, source TNO).

technical project risk based on subsurface key performance parameters, and to promote the bankability of projects. In this chapter, we describe the developed approach and its implications for resource potential. Detailed description of methodology and software tools are publicly available under the Dutch geothermal information system website Thermogis, www.thermogis.nl, and the website for data of the Dutch deep subsurface, www.nlog.nl.

The chapter is structured in four sections. In the first part, we outline the way aquifers have been selected for mapping and we explain the methodology for the geothermal characterization of aquifers, including depth and thickness, porosity, permeability, transmissivity, and underlying uncertainty. In the second part, we explain how we assess the thermal structure. In the third part, we present performance tools for resource assessment and prospect analysis, and finally we present an outlook to future development and challenges.

6.2. Geothermal Aquifer Characterization

Geothermal aquifer characterization is focused toward the assessment of the 3D structure of the aquifers in terms of depth and thickness, and transmissivity. Depth is critical as it is a prime indicator for production temperature, and transmissivity is a key performance indicator for successful geothermal exploration, since it is linearly proportional to the achievable flow rate in the aquifer as a function of pressure difference imposed on the injection and production wells at the reservoir level.

In this study, all known clastic aquifers in the subsurface known from hydrocarbon studies of the Netherlands in the Mesozoic and Permian sequence have been screened first. To be selected, an aquifer should be distributed over a large regional area of at least several tenths of km^2; the aquifer water temperature should be at least of 40°C. This corresponds to a minimum depth of 1000 m, assuming an average temperature gradient of 30°C/km and surface temperature of 10°C. This approach resulted in a selection of 20 aquifers (Pluymaekers *et al.*, 2012).

The existing detailed 3D subsurface models (Kombrink *et al.*, 2012; Duin *et al.*, 2006) (Figure 6.3), well data, and exploration and production studies developed for hydrocarbons (van Balen *et al.*, 2000; Duin *et al.*, 2006; Nelskamp *et al.*, 2012; Bruns *et al.*, 2015) include detailed structural geometrical skeleton of a limited main stratigraphic groups (Figure 6.3), but did not concern mapping of geothermal aquifer geometries and its properties.

Hydrocarbon studies have largely focused on valid trapping structures and assessment of hydrocarbon source maturation and migration (van Balen *et al.*, 2000), whereas assessment of flow properties has been secondary in importance. Further oil and gas wells and associated exploration studies have not been uniformly distributed throughout

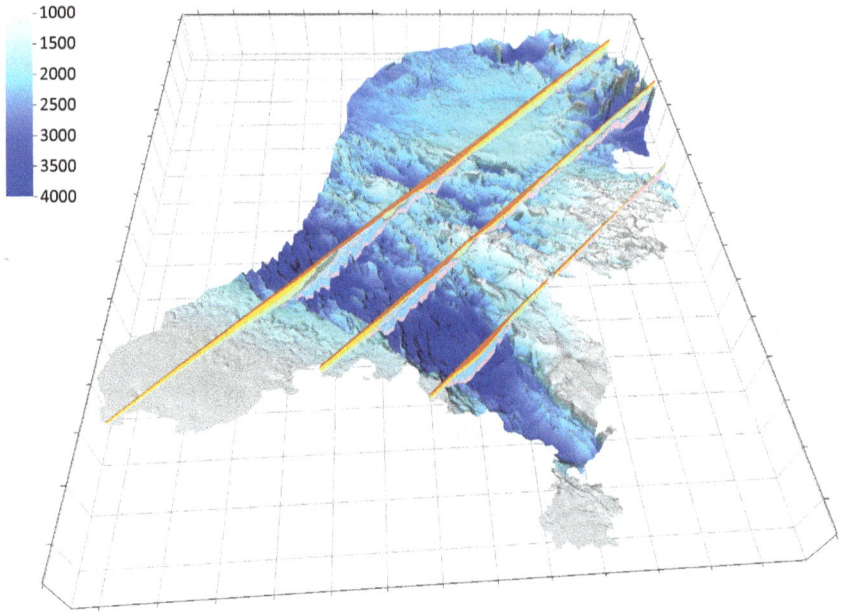

Figure 6.3   Bird's eye view of the base of one of the 10 mapped stratigraphic groups of the Mesozoic sedimentary sequence in the Netherlands (NLOG, 2015).

the area but preferentially located in structural high areas and in areas with proven hydrocarbon plays. In the depth domain, oil and gas exploration has largely focused toward specific reservoir levels, likely to trap oil and gas. As a consequence, stratigraphic levels suitable for geothermal heat production do only partially match laterally and stratigraphically with oil and gas reservoirs.

Consequently, crucial information of aquifers is missing both in terms of completeness in the mapped 3D structures as well as regional mapping of aquifer transmissivity. This has led us to develop a generic workflow, which is capable of producing a detailed 3D structure and flow properties (Figure 6.4). The workflow takes existing 3D subsurface models and core plug data and log data from wells as input. The process for mapping (maximum burial) depth and thickness of aquifers is presented first. The process for mapping transmissivity is denoted by five process steps, which are subsequently described is the sections below. The workflow is capable of predicting average values as well as underlying uncertainty.

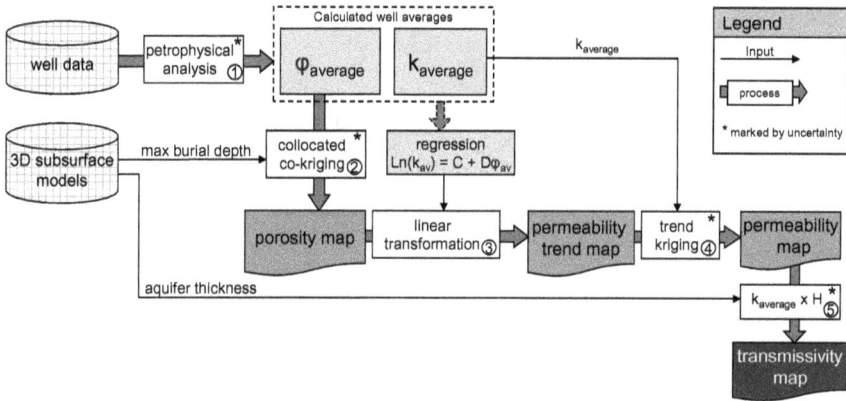

Figure 6.4   Process workflow for geothermal aquifer characterization.

The workflow steps (Figure 6.4) relies on a number of key assumptions, which are summarized prior to detailing the workflow:

1. For transmissivity, it is assumed that laterally aquifers are marked by no heterogeneity, in which case the transmissivity corresponds to the average permeability in the aquifer interval, multiplied by its thickness H. In our approach, the aquifer thickness H is the mapped thickness of the aquifer and is not further differentiated to net pay zones, through a cut-off value in permeability. Although it can be argued that average values of permeability would be higher in an adoption of net pay zones, the transmissivity of the aquifer will not change as the thickness is proportionally reduced.

2. It is assumed that the logarithm of permeability ($\ln k$) and the porosity (φ) holds a linear relationship. This relationship follows from progressively better flow performance with increasing pore space, and is supported by core data (Figure 6.5) and is widely used in basin and reservoir modeling (Hantschel and Kauerauf, 2009). Consequently, porosity, in absence of permeability, is a good indicator of permeability. The linear relationship is marked by considerable uncertainty as the relationship is strongly dependent on lithofacies and chemical cementation (Pape *et al.*, 1999).This type of linear relationship is used on a well basis to derive average permeability in the wells, and is also used to construct a predictive relationship for average permeability from average porosity of the aquifer.

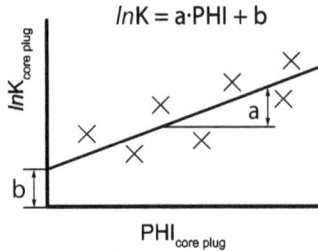

Figure 6.5   Linear relationship between porosity and the logarithm of permeability measurements observed in core plugs.

3. Commonly, it is assumed that porosity decreases with depth as a result of mechanical compaction. Porosity depth curves have been widely adopted in basin modeling to predict decrease of porosity with depth (Bond and Kominz, 1984; Allen and Allen, 2013) and hence permeability of sediments (Hantschel and Kauerauf, 2009). Burial depth should therefore be used in the spatial interpolation either through co-kriging or adopting a porosity–depth curve. Here we adopt data driven co-kriging, as porosity depth curves depend strongly on lithology and are marked by significant uncertainty (van Wees *et al.*, 2009; van Balen *et al.*, 2000), which is hard to constrain to the porosity data in the wells.

4. Mechanical compaction can be higher than expected for present-day burial if rocks have been buried more deeply in the past. The excess burial, or so-called burial anomaly, if significant (e.g., larger than a few hundred meters), and is taken into account in the spatial interpolation of porosity. Burial anomalies are caused by erosional events, bringing deeply buried rocks closer to the surface. However, if the erosion is followed by burial, which is equal in magnitude or higher than the erosion, then the burial anomaly is lost. In the Netherlands, burial anomalies can be significant over large regions (Nelskamp *et al.*, 2012; Pluymaekers *et al.*, 2012) and have been taken into account in porosity estimation through co-kriging with the reconstructed maximum burial depth instead of the observed burial.

6.2.1. Depth and Thickness

The depth and thickness of aquifers have been mapped by 2D geo-statistical interpolation techniques using the existing mapped major stratigraphic horizons of the onshore Netherlands (Duin *et al.*, 2006) as a geometrical skeleton. The stratigraphic horizons in this skeleton are limited to the main lithostratigraphic groups and lack information on aquifers. Further, this model has not been based on all available wells and 3D seismic as some of these data were not yet freely accessible during mapping.

5. The depth and thickness of the aquifers have been interpolated in the geometrical skeleton framework. In general, the aquifers have been marked by relatively uniform depositional circumstances, allowing a geostatistical approach to generate depth and thickness maps. The extent of the aquifer is taken from the geometrical skeleton model, based on wells penetrating the aquifer. The geometrical skeleton is leading in the structural interpretation, and the geostatistical interpreted thickness is truncated against the skeleton borders resulting in, mostly structurally controlled, jumps in thickness.

6. In the mapping of the depth of the aquifers, uncertainty has not been taken into account. This uncertainty consists of uncertainty in seismic interpretation, velocity modeling, and most important the data density of available (3D) seismic and well-data. The uncertainty is typically in the range of 5–10% and can be considered minor. Uncertainty in thickness has been modeled based on the geostatistical analyses of thickness of the aquifers in the wells only. It is assumed that this uncertainty is sufficient for capturing large wavelength effects.

6.2.2. Transmissivity

The first step in mapping the transmissivity is the determination of average porosity and the average permeability, as input for the generation of average porosity maps and as input for a predictive trend of average porosity and permeability respectively (Figure 6.4, process 1). To obtain average porosity and permeability, an extensive dataset of approximately 12,000 porosity and permeability core plug values and over 650 logs of onshore wells were used and corresponds to about 50% of theoretical availability of data. Complementary to digital available data, compiled results from reports on petrophysical analysis from various sources including public geothermal studies have been adopted.

6.2.2.1. Porosity

For each aquifer, the average porosity $\varphi_{average}$ was calculated for all wells with available digital porosity logs and core plug data. Standard oil and gas petrophysical workflow was used for porosity determination, from bulk density and neutron logs or both (Serra *et al.*, 1984). If core plug measurements were available, the calculated porosity logs were shifted to fit the core-porosity measurements. The logs have been arithmetic averaged to average aquifer porosity. When only core measurements were available, the average aquifer porosity has been assumed equal to the average of the core values.

6.2.2.2. Permeability

Permeability is directly derived from porosity in the wells. For this conversion, a linear relation between porosity and the logarithm of the permeability from core plug measurements has been used. This relation was determined by a least-square regression on the core plug measurements (Figure 6.5). For each aquifer interval, calculated porosity logs could be transformed to permeability logs if core plug measurements were available for the specific aquifer interval at the specific well. As core plug measurements are not available in each well, this resulted in a much smaller permeability dataset compared to the porosity dataset. The logs have been arithmetic averaged to an average aquifer (horizontal) permeability $k_{average}$. When only core measurements have been available, the average permeability has been assumed equal to the average of the core values.

6.2.2.3. Average porosity — average permeability trend

In a final step, the trend of average porosity and average permeability has been determined to be used in Section 6.3.3 to convert a porosity map into trend permeability map using the following equation:

$$k_{average} = C + D\varphi_{average} \qquad (6.1)$$

where C and D are the intercept and slope of the trend line of average porosity and the logarithm of average permeability. In a later process, a correction was made to account for the deviations of the calculated averaged permeability from the trend line, and the observed permeability at well locations. Figure 6.6 shows an example of the trend line for the Rotliegend aquifers, separating the trend lines, which have derived from wells where log and core have been available and the trend line derived from core analysis only. Figure 6.6 clearly demonstrates that using only core data bears a risk that high-permeable streaks, which weigh heavily in the average permeability, are missed and result in a permeability prediction, which is too low, as reflected

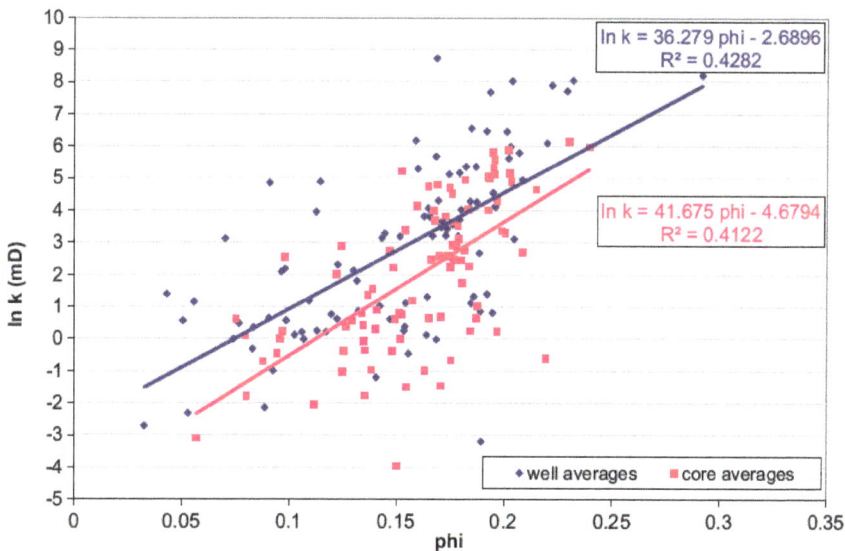

Figure 6.6   Average porosity and permeability from core plugs (pink) and logs (blue) for the Rotliegendes aquifer.

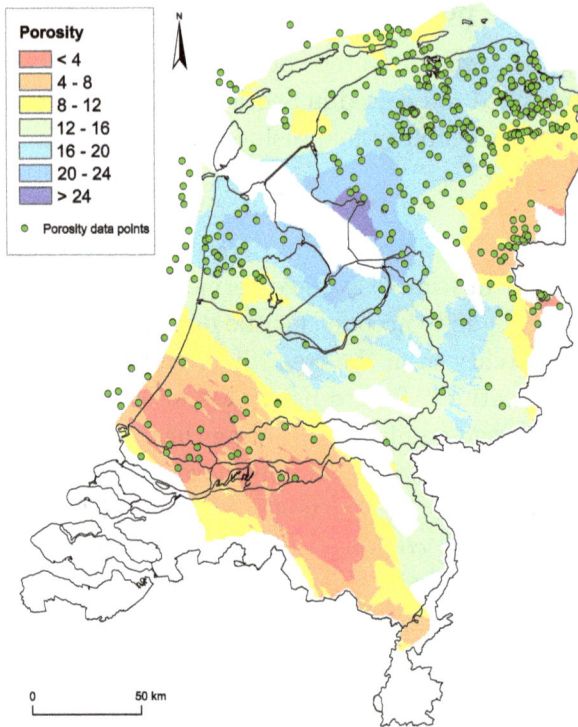

Figure 6.7 Average porosity map based on co-kriging with depth for the Rotliegendes aquifer group.

by the intercept value. However, the average core porosity values appear to agree well with the averages from the logs. For this reason, the core data in the workflow have been used to determine average porosity, but have not been used to determine average permeability and the trend line, when sufficient log data was available.

6.2.3. Porosity Maps

As mentioned before, the map view interpolation of the average porosity was done with the collocated co-kriging method, in which the burial depth as a second, collocated, variable has been taken into account (Figure 6.3, process 2).

6.2.4. Permeability and Transmissivity Maps

The average permeability map is constructed in two steps. First, the average porosity map is converted to a permeability trend map using Equation (6.2) (Figure 6.3, process 3). Subsequently, the permeability map is obtained from trend kriging the (limited number) $k_{average}$ data points were the permeability trend map was used as trend input (Figure 6.3, process 4). The resulting permeability map honors the data points, and deviates toward the permeability trend when no permeability data is available.

The transmissivity was calculated by multiplying the aquifer thickness with the obtained permeability (Figure 6.3, process 5). For uncertainty assessment, the following assumptions have been made:

1. The average porosity (Figure 6.7) has been assumed to be marked by a normal distribution with mean $\varphi_{average}$.
2. The standard deviation of normal distribution $(\varphi_{average})_{sd}$ at the well location is assumed to be approximately 5% and 10% of the mean for log and core derived values, respectively. For wells, average values have been extracted from literature, which were based on full petrophysical analysis $(\varphi_{average})_{sd}$ was assumed to be 3% of the mean value.
3. In between wells $(\varphi_{average})_{sd}$ is expressed as the standard deviation from the kriging interpolation. The used variogram is leading; the uncertainty at the well position is incorporated in the variogram (as a threshold value for the nugget).
4. The average permeability is assumed to be log normal distributed with average $\ln(k_{average})$ and standard deviation $\ln(k_{average})_{sd}$. is given by the standard deviation of the permeability trend (expressed as function of $(\varphi_{average})_{sd}$ cf. Equation (6.2)) and the standard deviation of the permeability kriging $\ln(k_{average})_{kriging-sd}$:

$$\ln(k_{average})_{sd} = \sqrt{(D(\varphi_{average})_{sd})^2 + \ln(k_{average})^2_{kriging-sd}} \quad (6.2)$$

The transmissivity distribution (Figure 6.8) has been generated through Monte Carlo sampling, assuming a normal distribution for H. From the Monte Carlo samples, the P10, P30, P50, P70, and P90

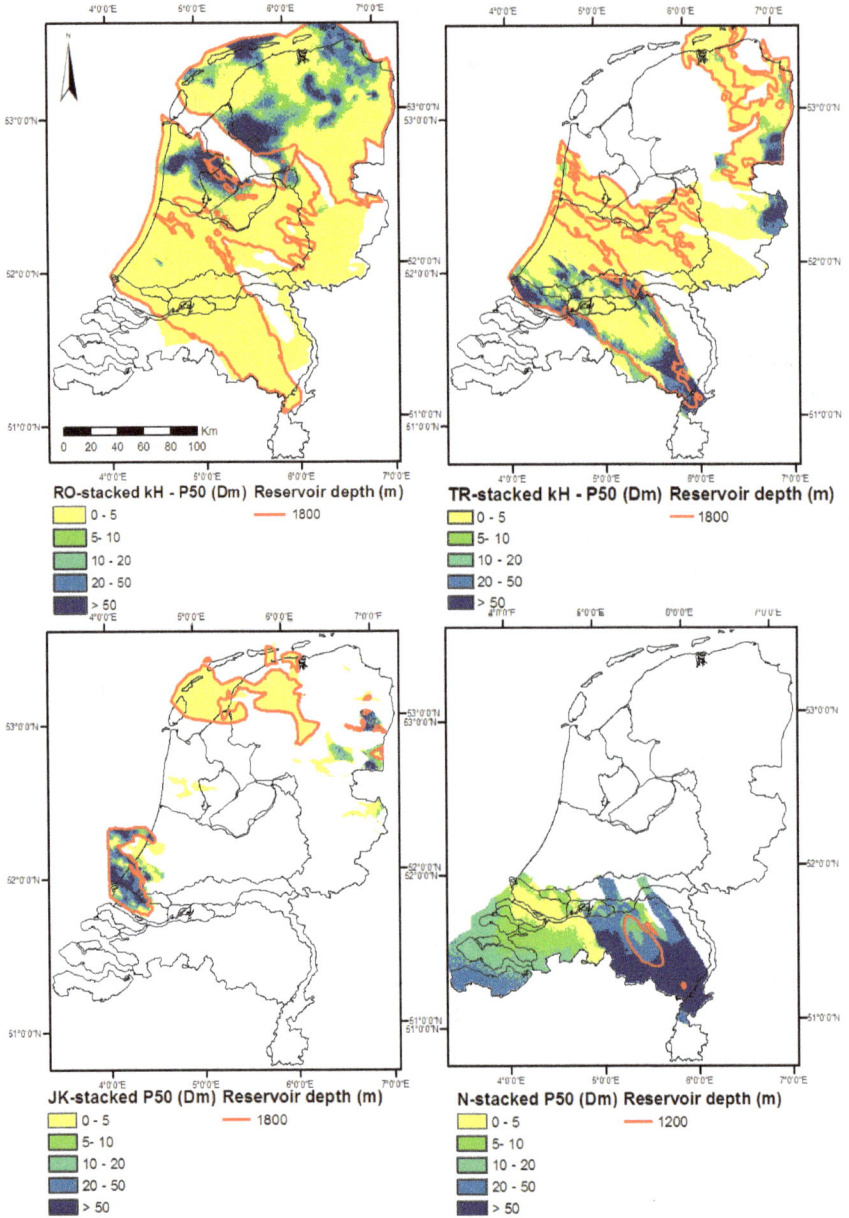

Figure 6.8   Transmissivity of the stacked sequences of: (a) Rotliegend; (b) Triassic; (c) Lower Cretaceous and Upper Jurassic; and (d) North Sea Supergroup. In (a), (b), and (c), areas shallower than 1200 m are not shown.

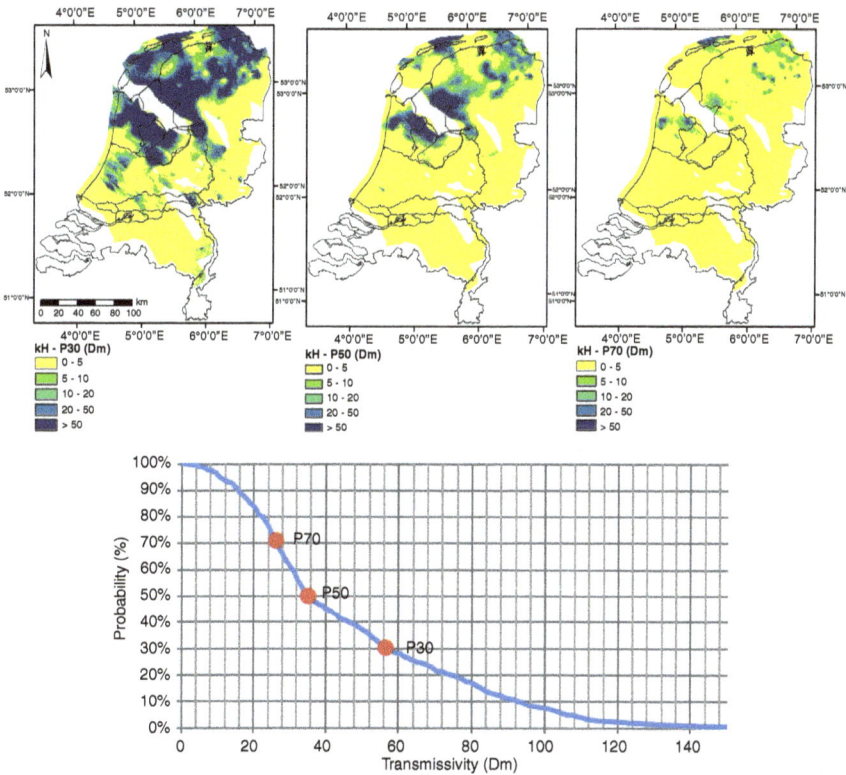

Figure 6.9 Probability distribution of the transmissivity of the Rotliegend aquifer, expressed in P30 (a) P50 (b) and P70 (c) maps. The expectation curve based Monte Carlo sampling is shown for a specific location.

values have been extracted at each location and have been used for map results in ThermoGIS (Kramers *et al.*, 2012). The modal P50 values of permeability do correspond to the permeability results of Equation (6.4). The map difference of transmissivity from P90 to P50 and the expectation curve at a particular location (Figure 6.9, middle panel) typically shows order of magnitude variations, demonstrating the dominant impact of uncertainty in permeability on transmissivity and associated performance.

6.3. Thermal Model

In the onshore hydrocarbon wells, 1241 measurements of temperature have been compiled in 437 wells to be used to build a 3D thermal

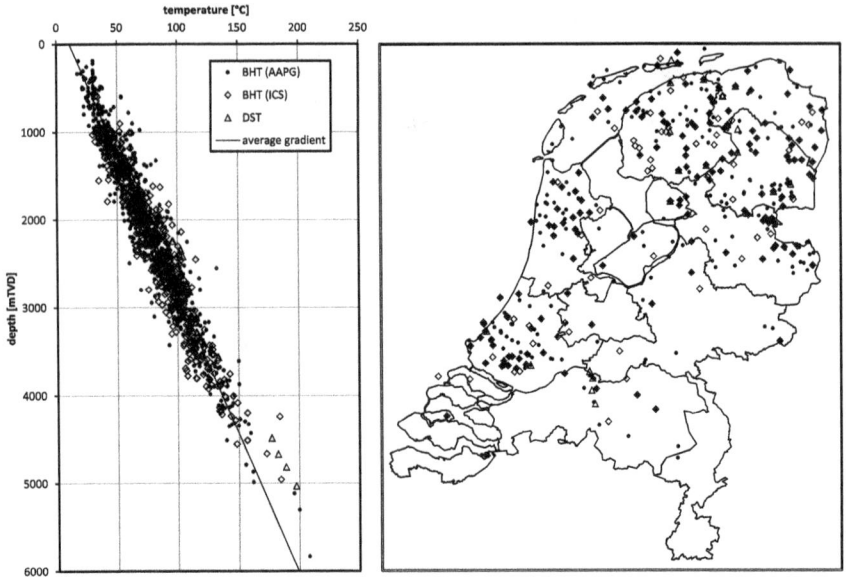

Figure 6.10 (Left) Corrected temperature data; (right) spatial distribution of temperature data.

model (Bonté *et al.*, 2012). Of this, 98.8% (1189) of the available measurements are bottom hole temperature (BHT) taken during drilling and need to be corrected due to the thermal perturbation created by mud circulation. The remaining 1.2% (52) is composed of Drill String Test (DST) measurements, with a value close to the formation (i.e., $\pm 5°C$). In these, 412 BHT measurements are highly reliable values, which could be corrected with the Instantaneous Cylinder Source (ICS) method (Goutorbe *et al.*. 2007) and the remaining 829 BHT values with a lower reliability have been corrected with the American Association of Petroleum Geologists (AAPG) method (Bonté *et al.*, 2012), calibrated by the ICS-corrected measurements. The average thermal gradient of this whole data set is $31°C$ km^{-1} with a surface temperature of $10°C$ (Figure 6.10).

For the construction of a thermal model from the temperature data set, we deployed a forward 3D thermal modeling approach using a tectonic heat flow method, which is based on the thermal properties of the sedimentary structure (Bonté *et al.*, 2012). In the model,

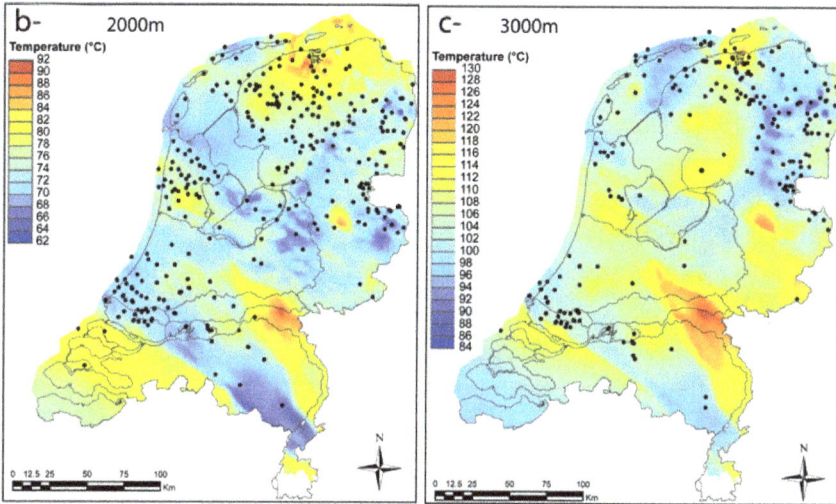

Figure 6.11   Depth slices of 3D forward model of temperature calibrated to well data, at 2000 m and 3000 m depth.

transient thermal effects of sedimentation and erosion have been taken into consideration, as well as surface temperature changes over the past 20 million years. The basal heat flow of the sediments in the model is constrained by lithospheric thickness and radiogenic heat production in the crust. A best-fit model for variation in lithospheric thickness and variability in radiogenic heat production in the crust has been constructed from a iterative calibration to well data. The resulting 3D thermal model allows to present both isodepth maps, temperature on profiles, and geological interfaces such as the mid-depth of the geothermal aquifers (Bonté *et al.*, 2012). Figure 6.11 shows the predicted temperatures at a depth of 2000 and 3000 m.

6.4. Resource Potential

A key question in targeting aquifers is which subsurface conditions, and underlying uncertainties, are key to performance. Understanding this relationship allows to focus regional mapping and data compilation efforts in geothermal exploration, especially if it is possible to building from a wealth of existing oil and gas data in mature basin areas such as the Netherlands. In this section, we highlight the importance of depth-dependent transmissivity and temperature trends in aquifers in the Netherlands, derived from oil and gas data, for performance characteristics.

The performance of a geothermal doublet in terms of power produced E is linearly proportional with the temperature difference ΔT of the produced and reinjected temperature of the percolating brine and the achievable flow-rate Q:

$$E = \Delta T \, Q \, c \qquad (6.3)$$

In principle, the performance prediction of a geothermal doublet at a given site is to be evaluated by a reservoir model acting as a base for a reservoir simulation. Reservoir simulations facilitate to predict the mass and volume flow of the brine, the energy extraction, the temperature, and pressure evolution as well as the lifetime of the reservoir in question in a probabilistic manner. Such an approach is confined by two aspects: (1) Applying a reservoir simulation is a complex procedure and requires considerable resources for purchasing and maintaining corresponding software codes and (2) sufficient data to set up a reliable conceptual reservoir model are in most cases not available at the initial state of geothermal resource assessment. To overcome these restrictions, a simplified methodology has been developed for performance calculations of doublet systems, and has been made publicly available as a software tool called DoubletCalc (van Wees *et al.*, 2012). This compact and easily usable tool allows a probabilistic based site-specific performance prediction of a geothermal doublet based on a moderate set of geotechnical input parameters. The software and the source code are available in the public domain at www.nlog.nl.

The methodology of DoubletCalc follows the logic of a so-called fast model (van Wees *et al.*, 2010, 2012), which is capable of predicting within a matter of seconds technical and economic performance indicators, under simplified model assumptions for reservoir behavior, cost engineering, and economics. Further, through a Monte Carlo sampling approach, the fast model, is capable of analyzing the effect of uncertainties in the subsurface on performance characteristics. Doing so, DoubletCalc is capable of modeling a doublet's behavior and underlying uncertainty, sufficiently accurate for a performance assessment at an early stage of exploration.

6.4.1. Sensitivity Analysis

The fast model calculations provide an effective way to assess sensitivity of performance as a function of variability of technical input parameters. This can facilitate in understanding sensitivity of performance to reservoir parameters and provides scope for optimizing engineering parameters. Figure 6.12 depicts the sensitivity of the performance of a 100 m thick aquifer, for the average permeability depth trend of the Rotliegend aquifer, a thermal gradient in accordance with the average gradient observed in the temperature data presented in Figure 6.10 and engineering and economic parameters outlined by van Wees *et al.* (2012). The figure clearly highlights a strong reduction of transmissivity with depth, which is expected from the reduction of porosity with depth. At high transmissivity values, doublet power increases, due to increasing temperature and associated ΔT. However, a threshold transmissivity, the trade-off between increase in temperature and reduction of transmissivity results in lower power values, resulting in theoretical optimum depth between 1.5 and 2 km in terms of maximum power produced at levelized costs of energy (LCOE). This applies for the parameters considered here. One should bear in mind that actual reservoir parameters can considerably vary spatially and are subject to major uncertainty, which would result in different predictions. Furthermore, development scenarios and engineering design has a strong influence on the outcome.

6.4.2. Probability of Success and Insurance Schemes

The stochastic performance computation is conducted by Monte Carlo simulations. The number of simulations can be chosen by the user. About 500–1000 simulations represent a setting delivering sufficient data for a reliable uncertainty assessment, for the uncertainty in input parameters as described in the previous section while requiring moderate calculation times.

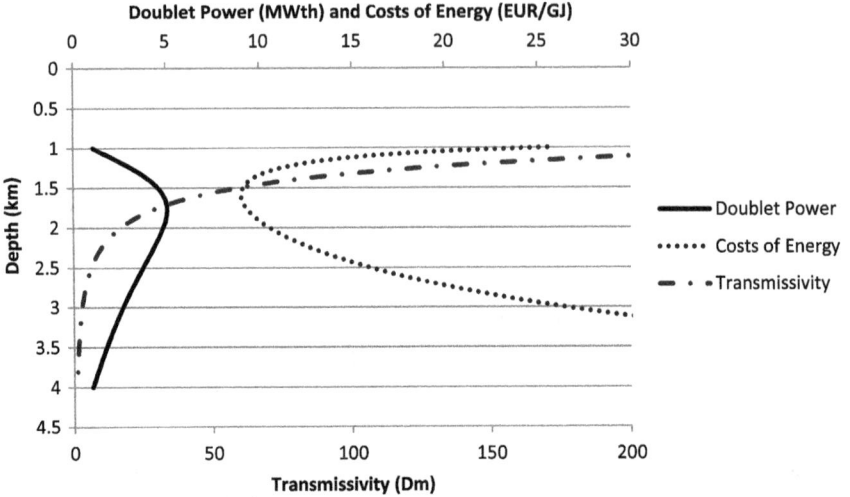

Figure 6.12 Doublet power and LCOE modeled with DoubletCalc for an aquifer in the Rotliegendes group.

The stochastic output can be viewed in a tabulated form or stochastic screen plots for selected economic and geotechnical input and output parameters. In particular, a cumulative density — or so-called expectation plot — of predicted power (Figure 6.12) can be used for assessing probability of success (POS) to comply to a Dutch Insurance Fund to recover costs of exploration drilling. Within the scope of this insurance scheme, the POS is determined through the probability of predicted power production to exceed a given minimum value, dictated by the business case of the project developer. This probability can be easily determined from the expectation plot as the vertical coordinate of the expectation curve, where its x coordinate corresponds to the minimum power requirement (Figure 6.13). In the business cases for project developers, the threshold power should be financially underpinned by their proprietary cash flow calculations and underlying constraints.

6.4.3. Prospectivity Maps

In the regional resource assessment, two aims are served. The first, presented in this section, is to obtain a map-based overview of

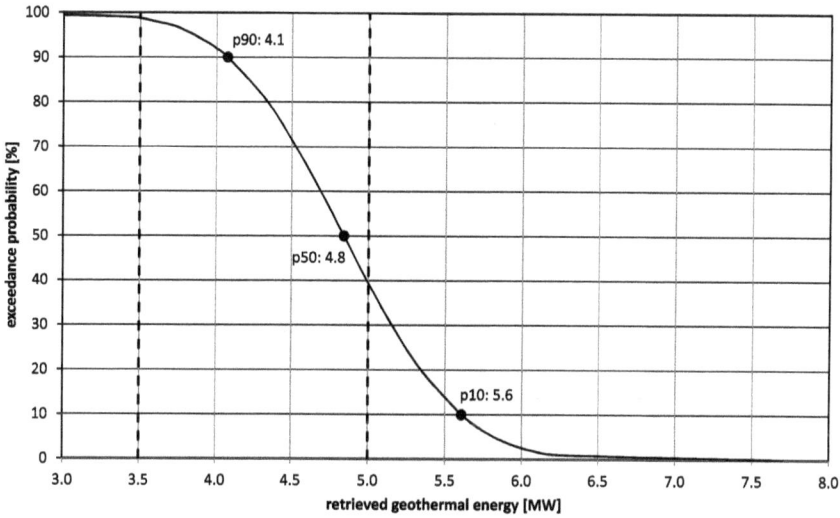

Figure 6.13   Expectation curve of doublet power predicted by DoubletCalc. Left and right dashed lines represent minimum required and desired power of the business case respectively. The expectation curve is marked by very low (ca. 1%) probability that the project will not reach the minimum required power.

prospective regions for geothermal direct heat applications. The second is to obtain an estimate of total amount of geothermal energy, which can be extracted from the subsurface, and will be presented in Section 6.4.4.

For both assessments, we aimed to take into account the effect of uncertainties, in order to be able to indicate not only most likely but also possible estimates with lower probability. This is common practice in resource reporting in the oil and gas industry (Etherington and Ritter, 2007). To this end, doublet performance calculations have been performed on geological properties of the aquifers for P30 (less likely), and most likely (P50) values of transmissivity. For the performance computation, the uncertainties in temperature, depth and well engineering properties were neglected. The argument to neglect these effects is that their influence on the performance is an order of magnitude lower compared to the uncertainty in transmissivity (see also Pluymaekers *et al.*, 2012).

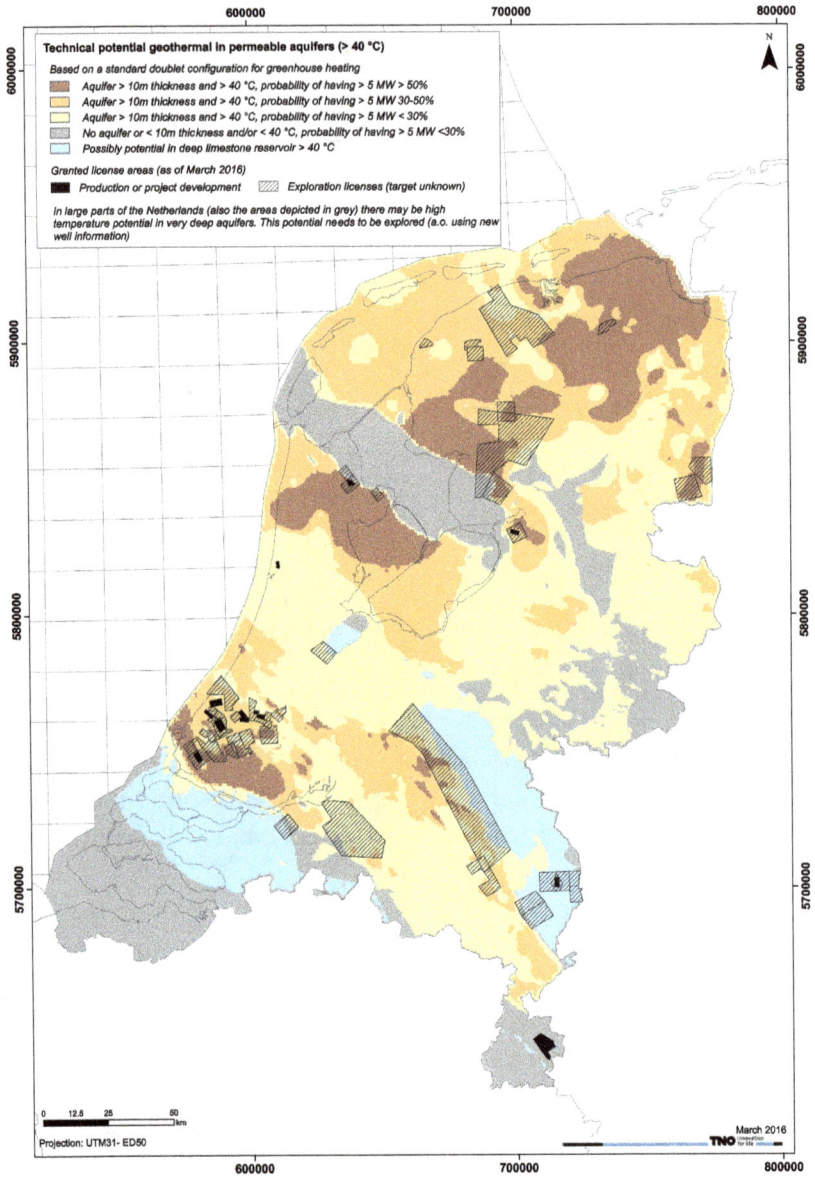

Figure 6.14   Prospectivity map of deep subsurface aquifers, overlain with exploration and production licenses.

For prospectivity, a major aim has been to present an indicative expected power map, which can be retrieved by doublets. This is a technical map, not taking into account economic constraints for power production. However, we enforced a minimum for coefficient of performance (COP) of 15, which is defined as the ratio of the thermal power produced and power consumption of the pumps for driving the thermal loop. The map is differentiated in classes based on the P30 and P50 power values. The compilation takes the P50 map as input, replacing P50 values with P30 values when predicted power is lower than 5MW$_{th}$. The resulting map has been subdivided into three potential classes.

- Good: power potential >5 MW$_{th}$, with ≥50% probability
- Possible: power potential >5 MW$_{th}$, with ≥30% probability
- Aquifers >10 m thickness, depth > 1000 m

The last class corresponds to presence of suitable aquifers, however there is less than 30% probability that aquifers that they are capable to produce >5 MW$_{th}$. The resulting map (Figure 6.14) is indicative and should be interpreted with care. More detailed local studies are required to enhance the regional picture. However the overall pattern of predicted favorability for prospectivity fits well with the distribution of exploration licenses and producing plants.

6.4.4. Resource Potential

To assess a cumulative resource estimate, in recent proposals for geothermal resource assessment (Edenhofer *et al.*, 2011, Beardsmore *et al.*, 2010), generally a distinction is made between theoretical potential describing the total amount of heat in the subsurface (aquifers), which can be used for a particular application, and technical potential, which forecasts the amount of heat that can be really extracted. Following Beardsmore *et al.* (2010), technical potential is expressed as a yearly extractable quantity [PJ km^{-2} yr^{-1}], adopting a lifetime of 30 years for harnessing the earth's heat in production. Such a representation allows to match spatial variability and magnitudes of technical potential with density in heat demand and is highly valuable for policymaking purposes. The technical potential is dependent on technical and economic factors. We have chosen to differentiate the technical potential into a theoretical upper bound (technical potential) and a lower bound (recoverable heat), which are respectively unconstrained and constrained by project economics.

For a national assessment, first the theoretical potential for aquifers is calculated and subsequently corrected for a technical ultimate recovery (UR) factor (technical potential) and economic constraints in heat extraction (recoverable heat). The calculation of technical potential involves a simple volumetric calculation of heat in place corrected for application-specific constraints of (temperature-dependent) conversion efficiency. The details of conversion efficiency, which are application dependent, are described in more detail in Kramers *et al.* (2012) for greenhouse and spatial heating applications. This step also takes into account the UR. For the UR, it is assumed that due to legal reasons, doublets are oriented in rectangles, enclosed by circles, which are centered at the injector and producer well at reservoir level and which touch each other at half way. In a doublet system of an aquifer, about 50% of heat enclosed in the rectangle can be technically recovered, before thermal breakthrough (cf. Gringarten, 1978). Further the layout of rectangle of multiple doublets will not be ideal, due to geological and economic reasons, leaving unrecovered heat in the space, which cannot be filled in further by new doublets. Taking into account these effects, it is assumed that the UR is about 33% of the HIP corrected for conversion efficiency.

The resulting technical potential estimate can be subsequently filtered to economic constraints to obtain recoverable heat. This approach incorporates the economics of the performance calculations detailed in van Wees *et al.* (2012) (Figure 6.15). In this process, the volume of rock representing recoverable heat is a subset of the technical potential, limited to the region where LCOE is lower than a threshold value. An estimate of the national recoverable heat corresponds to an aerial summation of the P50 map.

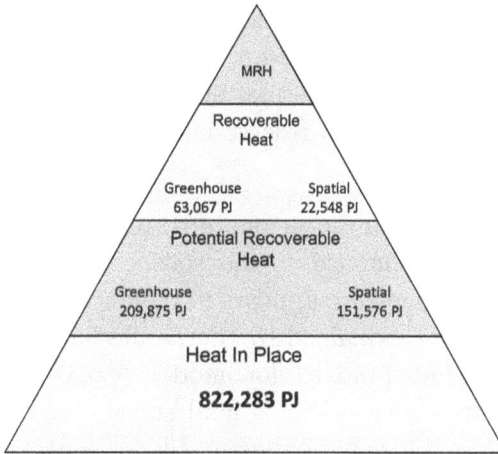

Figure 6.15   Filtering pyramid showing the total heat content for RH (recoverable heat), PRH (potential recoverable heat) and HIP (heat in place).

6.5. Production Challenges and Future Outlook

The resource and performance assessment methodology has been developed to designate prospective high-permeable clastic aquifers and to assess the amount of potential geothermal energy in the Netherlands. It highlights the geothermal potential through an overview of the expected doublet power, which can be retrieved and a web-based portal ThermoGIS (www.thermogis.nl) (Kramers *et al.*, 2012), allows to assess all underlying data including maps of depth, thickness, permeability, and temperature of the aquifers. Public access to subsurface information, dedicated to geothermal exploration, and public tools such as DoubletCalc contributed significantly to geothermal exploration in the Netherlands.

In the fast growth of Dutch geothermal plants, geothermal production has been confronted with a range of challenges and concerns to be addressed. These include worse than expected injectivity, assessment of land subsidence, the scope for reducing injection temperature for enhanced performance, prediction and mitigation of mineral scaling, and dealing with oil and (CO_2-containing) gas co-production.

In the recent years, various research studies, in close cooperation with project developers, provided a unique opportunity to study in detail data of the major geothermal projects, providing us with case studies for the different target reservoirs in the Netherlands, shedding light on major issues:

- For injectivity, many of the encountered problems most likely were related to near-well damage to the wells.
- Lowering the temperature appears to enhance injectivity in many cases. This may be explained by thermomechanical effects, which are not included in standard flow models (Loeve *et al.*, 2015).
- Another geomechanical effect is land subsidence due to reservoir cooling. We found it may be measurable for some geothermal systems but generally not significant (Fokker *et al.*, 2015).
- A geochemical assessment of a degassing loop for geothermal water has been performed, to assess the sensitivity for scaling. With dissolved CO_2 present in the reservoir, the scaling potential is highest due to degassing — if CO_2 is kept in solution, this risk is largely reduced. The predicted amount of scaling is sensitive to the location-specific chemistry and distinct differences exist for the different studied reservoir types.

Facing the challenges provided a valuable learning curve for geothermal development in the Netherlands and clastic aquifer production in general, augmenting to reliability of the performance of the doublets. In addition, the resource base can be extended through enhanced reservoir and thermal loop engineering, including heat pumps. In particular, hydraulic stimulation or alternative well designs leading to higher productivity for the wells allow to extend exploration and production into deep depth levels ($\gg$2000 m) where permeability deteriorates as a consequence of mechanical compaction and cementation (Pluymaekers *et al.*, 2015). The implications on subsurface potential are significant as it unlocks considerable clastic aquifer potential. Furthermore, synergy with oil and gas production activity can be enhanced through a double play perspective in exploration (van Wees *et al.*, 2015), allowing to share risks and drilling costs for exploration.

References

Allen, P.A. and Allen, J.R. 2013, *Basin Analysis: Principles and Application to Petroleum Play Assessment*, Chichester: Wiley.

Beardsmore, G.R., Rybach, L., Blackwell, D., and Baron, C. 2010, A protocol for estimating and mapping global EGS potential, *GRC Transactions*, 34, 301–312.

Bond, G.C. and Kominz, M.A. 1984, Construction of tectonic subsidence curves for the early Paleozoic miogeocline, southern Canadian Rocky Mountains: Implications for subsidence mechanisms, age of breakup, and crustal thinning, *Geological Society of America Bulletin*, 95(2), 155–173.

Bonté, D., van Wees, J.D., and Verweij, J. 2012, Subsurface temperature of the onshore Netherlands: new temperature dataset and modelling, *Netherlands Journal of Geosciences*, 91(4), 491–515.

Bruns, B., Littke, R., Gasparik, M., Wees, J., and Nelskamp, S. 2015, Thermal evolution and shale gas potential estimation of the Wealden and Posidonia Shale in NW-Germany and the Netherlands: a 3D basin modelling study, *Basin Research*, Published in 2016. Basin Research 28, 2–33, doi: 10.1111/bre.12096

Duin, E.J.T., Doornenbal, J.C., Rijkers, R.H.B., Verbeek, J.W., and Wong, T.E. 2006, Subsurface structure of the Netherlands — results of recent onshore and offshore mapping, *Netherlands Journal of Geosciences*, 85(4), 245.

Edenhofer, O., Pichs-Madruga, R., Sokona, Y., Seyboth, K., Kadner, S., Zwickel, T., Eickemeier, P., Hansen, G., Schlömer, S., and von Stechow, C. 2011, *Renewable Energy Sources and Climate Change Mitigation: Special Report of the Intergovernmental Panel on Climate Change*, Cambridge University Press, Cambridge.

Etherington, J.R. and Ritter, J.E. 2007, The 2007 SPE/AAPG/WPC/SPEE Reserves and Resources Classification, Definitions and Guidelines. Defining the Standard!, *Hydrocarbon Economics and Evaluation Symposium*. Society of Petroleum Engineers.

Fokker, P.A., van der Veer, E.F., and van Wees, J.D. 2015, Surface movement induced by a geothermal well doublet, World Geothermal Congress 2015 Melbourne.

Goutorbe, B., Lucazeau, F., and Bonneville, A. 2007, Comparison of several BHT correction methods: a case study on an Australian data set, *Geophysical Journal International*, 170(2), 913–922.

Gringarten, A.C., 1978. Reservoir lifetime and heat-recovery factor in geothermal aquifers used for urban heating, *Pure And Applied Geophysics*, 117, 297–308.

Hantschel, T. and Kauerauf, A.I. 2009, *Fundamentals of Basin and Petroleum Systems Modeling*, Springer Science & Business Media, Berlin.

Kombrink, H., Doornenbal, J., Duin, E., Den Dulk, M., ten Veen, J., and Witmans, N. 2012, New insights into the geological structure of the Netherlands; results of a detailed mapping project, *Netherlands Journal of Geosciences*, 91(4), 419–446.

Kramers, L., van Wees, J.D., Pluymaekers, M., Kronimus, A., and Boxem, T. 2012, Direct heat resource assessment and subsurface information systems for geothermal aquifers; the Dutch perspective, *Netherlands Journal of Geosciences,* 91(4), 637–649.

Loeve, D., Veldkamp, J.G., Peters, E., and van Wees, J.D. 2015, Development of Thermal Fractures in Two Dutch Geothermal Doublets, The Third Sustainable Earth Sciences (SES) Conference & Exhibition, Celle, Germany.

Nelskamp, S., van Wees, J.D., and Littke, R. 2012, Structural Evolution Temperature and Maturity of Sedimentary Basins in the Netherlands: Results of Combined Structural and Thermal Two-Dimensional Modeling.

Pape, H., Clauser, C., and Iffland, J. 1999, Permeability prediction based on fractal pore-space geometry, *Geophysics,* 64(5), 1447–1460.

Pluymaekers, M., Veldkamp, J.G., and van Wees, J.D. 2015, A Generic Workflow to Assess (Stimulated) Clastic Aquifer Potential, World Geothermal Congress 2015 Melbourne.

Pluymaekers, M., Kramers, L., van Wees, J.D., Kronimus, A., Nelskamp, S., Boxem, T., and Bonté, D. 2012, Reservoir characterisation of aquifers for direct heat production: Methodology and screening of the potential reservoirs for the Netherlands, *Netherlands Journal of Geosciences,* 91(4), 621–636.

Serra, O., Westaway, P., and Abbott, H. 1984, *Fundamentals of Well-Log Interpretation,* Elsevier, Amsterdam.

van Balen, R.T., van Bergen, F., Leeuw, C.d., Pagnier, H., Simmelink, H., van Wees, J.D., and Verweij, J.M. 2000, Modelling the hydrocarbon generation and migration in the West Netherlands Basin, the Netherlands, *Netherlands Journal of Geosciences,* 79(1), p. 29.

van Wees, J.D., Bonté, D., Bertani, R., and Genter, A. 2010, Using Decision Support Models to Analyse the Performance of EGS Systems, Proceedings of the World Geothermal Congress, 2010 Bali, Indonesia, pp. 25–29.

van Wees, J.D., Kramers, L., Mijnlieff, H., De Jong, S., and Scheffers, B. 2015, Geothermal and Hydrocarbon Exploration: the Double Play Synergy, World Geothermal Congress 2015 Melbourne.

van Wees, J.D., Kronimus, A., van Putten, M., Pluymaekers, M., Mijnlieff, H., van Hooff, P., Obdam, A., and Kramers, L. 2012, Geothermal aquifer performance assessment for direct heat production–Methodology and application to Rotliegend aquifers, *Netherlands Journal of Geosciences,* 91(4), 651–665.

van Wees, J.D., van Bergen, F., David, P., Nepveu, M., Beekman, F., Cloetingh, S., and Bonté, D. 2009, Probabilistic tectonic heat flow modeling for basin maturation: Assessment method and applications, *Marine and Petroleum Geology,* 26(4), 536–551.

Chapter 7

Geothermal Heat Pump Development: Trends and Achievements in Europe

Ladislaus Rybach

Institute of Geophysics, ETH Zurich, Switzerland
rybach@ig.erdw.ethz.ch

Burkhard Sanner

GeoTrainet AISBL, Brussels, Belgium
sanner@sanner-geo.de

Thanks to the regular reporting at World Geothermal Congresses, there is a reliable statistical base to assess the status and development trends in European countries. The assessment reveals that the geothermal heat pump (GHP) technology is present and growing in application. The top five countries engaged in developing this technology are Sweden, Germany, France, Switzerland, and Norway.

The main message is that there is a lively GHP scene in Europe; GHPs are the most powerful systems in geothermal direct use. However, the absolute numbers and growth rates are very different from country to country, even if allowance is made for the country size and/or population.

The GHP development and use is by no means uniform; there are leading countries (also by global standards) that have reached high areal GHP installation densities and per-capita energy uses of GHP systems already since many years, whereas others have little or even no use of this technology. It can be expected that with time, more and more countries will embark on this versatile and environmentally friendly technology.

Considerable efforts are still needed to disseminate the message about the technical feasibility and reliability, sustainability, economic advantages, and environmental benefits of GHP systems to developers, decision makers, and even to the general public. The European Geothermal Energy Council (EGEC) is highly instrumental with manifold activities to help further GHP development.

PART A: COUNTRY-WISE

7.1. Introduction

Geothermal heat pump (GHP) systems provide the largest contribution to global direct use (325 PJ/yr, 55.3% of total; Lund and Boyd, 2015). The GHP development speed, thanks to the increasing dissemination of the GHP technology — based on know-how transfer and many successful examples of application — is indeed impressive: over the years 2000–2014, the average annual growth rate is about 20%, by far the highest in all other geothermal energy utilization technologies. As Figure 7.1 shows, the global growth is somewhat slower since 2010; nevertheless the growth continues at still a high level.

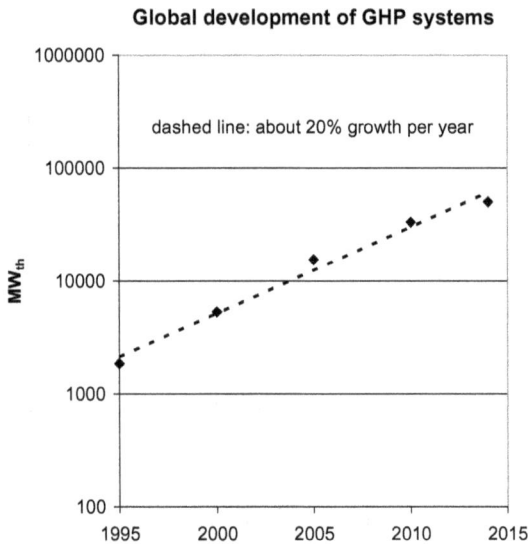

Global development of GHP systems

dashed line: about 20% growth per year

Figure 7.1   Global GHP system growth is nearly exponential over the time period 1995–2014 (data from Lund and Boyd, 2015).

For Europe, only one country reported numbers about GHP use at the World Geothermal Congress (WGC) 1995 whereas at WGC 2015, there have been 35 national reports with data about installed capacities and annual energy production. Obviously, all European countries will sooner or later use GHP systems; some of them to go from practically zero to quite somewhere.

The purpose of this compilation is to highlight the developments and achievements in GHP technology dissemination in Europe, which by all means is highly country-specific. Fortunately, since 1995, there is a World Geothermal Congress held every five years; for the Conference Proceedings, national country reports can be requested. For direct use including GHPs, these are delivered since WGC 2000 in a prescribed format. These reports provide the best statistical data on the international geothermal scene. John Lund (Geo-Heat Center, Oregon Institute of Technology) is highly instrumental in developing the reporting format, in selecting capable and motivating willing national reporters and in compiling, presenting, and interpreting the results.

The data used to find and define development trends for European countries are taken from the four WGC direct use compilations: the summary papers of Lund and Freestone (2000) Lund *et al.* (2005, 2010) and of Lund and Boyd (2015). The actual data apply — due to the time needed for manuscript preparation of the national authors, compilation of results for the summary papers, etc. — to one or two years before the conferences; therefore, the WGC 2015 data compilations are usually generally based on 2013 data. The GHP status as reported at WGC 2015 is represented for countries with more than 100 MW_{th} installed capacities is given in Figure 7.2.

In the following, the data of the four WGCs are compiled country-wise. Then from this data base, development trends are extracted and made visible, and impressive examples of growth are highlighted. In addition, especially successful countries are identified, on the basis of a specific ranking procedure. The database consists of numbers about installed capacities (MW_{th}) and annual heat energy (TJ/yr). These are listed country-wise, as they can be found in the WGC 2000, 2005, 2010, and 2015 summary papers on direct use, in the appendix.

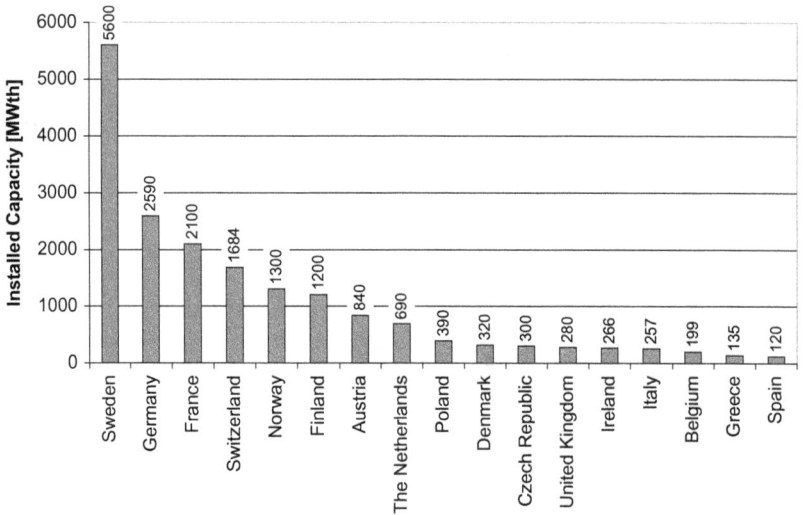

Figure 7.2   GHP capacities as reported at WGC (2015), for countries above 100 MW_{th}.

7.2. Development Trends in the European Countries

The most straightforward indicator of GHP technology application in a given country is the total installed GHP thermal capacity, expressed in thermal megawatts (MW_{th}). It is interesting to see and to compare the numbers from WGCs 2000, 2005, 2010, and 2015. Only one country (Ukraine) has absolutely no data; from this, it can be concluded that practically no GHPs are installed there today. For Armenia, estimates were only available at WGC 2005. No data are reported from 20 countries at WGC 2000, nine countries at WGC 2005, and five countries at WGC 2010.

Iceland represents a special case in GHP applications: whereas this country is clearly the world leader in geothermal district heating (DH), there are no ground-coupled heat pumps operating. The reason is that at the few locations outside of DH networks, the buildings use electric heaters, running on well-subsidized electricity; substantially more expensive GHP systems (extra drilling) are far beyond being economical.

There are several discrepancies in the data reported by the WGC direct use summary papers. These can partly also be attributed to the fact that not always the same people or institutions prepared the country update reports for the various WGCs. So, for example, a total installed GHP capacity of 48 MW_{th} was reported for France at WGC 2005 in contrast to only 16 MW_{th} at WGC 2010. Other contradicting examples are marked in the appendix.

Whereas some countries like Germany, Greece, Sweden, or Switzerland show continuous growth in installed capacity starting already with WGC 2000, other countries like France or Turkey come in later but show spectacular growth, especially Norway.

Today's top five countries with more than 1 GW_{th} installed capacities reported at WGC 2015 are Sweden (5.6 GW_{th}), Germany (2.6 GW_{th}), France (2.1 GW_{th}), Switzerland (1.7 G GW_{th}), and Norway (1.3 GW_{th}). Their development trends over the years are shown in Figure 7.3, which displays the data presented in the different WGC direct use summary papers. These countries have correspondingly high amounts of annually produced heat (TJ/yr). The greatest increases in the last five years in GHP utilization can be seen in Sweden, France, and Switzerland.

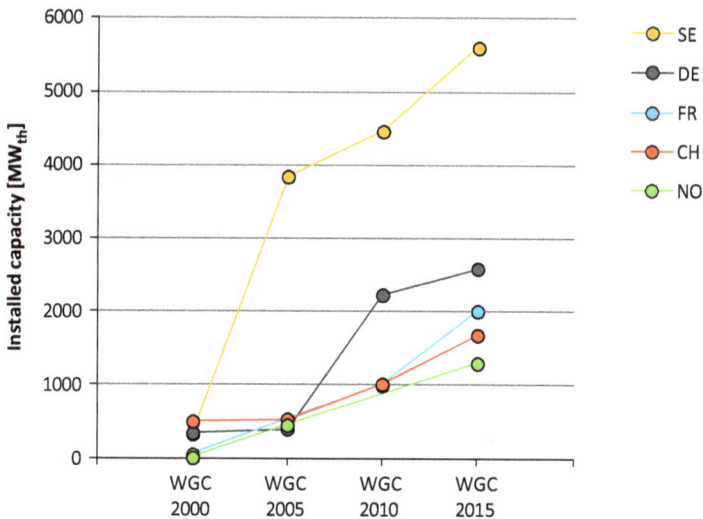

Figure 7.3 Development trends of installed GHP capacity in the top five countries in Europe. SE: Sweden, DE: Germany, FR: France, CH: Switzerland, NO: Norway. Data from the appendix.

7.3. Achievements and Awards: Development Ranking

The above mentioned numbers of installed capacity and annually produced heat are certainly impressive but do not reflect the actual achievements of the individual countries. For the latter, the country size (population, area) needs to be considered. To bring these characteristics to play, the so-called ranking criteria introduced by Lund *et al.* (2005) are applied:

- Installed capacity per country area,
- Installed capacity per capita,
- Heating energy per country area,
- Annually produced heat per capita.

Besides many traditionally smaller objects like single-family houses, numerous larger complexes with dozens of MW_{th} capacities are nowadays equipped with GHP systems. To take this into account, the concept of "equivalent 12 kW_{th} units" was conceived and introduced (Lund *et al.*, 2003); by this means a large unit with, for example, 12.6 MW_{th} counts now in the statistics as 1050 12 kW_{th} equivalent units. A capacity of 12 kW_{th} is needed for an average single-family house; the areal density of 12 kW_{th} units give a feeling about installation density in a given area or country. Thus the number of 12 kW_{th} equivalent units is an additional criterion, leading to the rankings in equivalent units per area. The latter gives impressive numbers like those for Switzerland.

All European countries with more than 100 MW_{th} have been considered for ranking. Only 17 countries comply with this criterion. In calculating the ranking criteria for the individual countries, a specific database has been assembled (Table 7.1). It contains only the countries considered.

On the basis of the calculated ranking numbers is Table 7.1, the top five countries (five highest numbers in the ranking categories) have been selected. Table 7.2 presents the rankings along with the calculated numbers.

Figures 7.4–7.8 visualize the ranking results, along with the numbers calculated.

Table 7.1   Database for country ranking.

Country	Area (km^2)	Population (million)	MW$_{th}$	TJ/yr	Number of 12 kWt equ. units	Equ. units/ area	MW$_{th}$/ area	W$_{th}$ per capita	TJ/yr/ area	MJ/yr per capita
Austria	83871	8.22	840	4990	70000	0.835	0.010	102.2	0.059	607
Belgium	30528	11.24	199	756	16583	0.543	0.006	17.7	0.005	67
Czech Republic	78867	10.54	300	1700	25000	0.317	0.004	28.5	0.021	161
Denmark	43094	5.57	320	3400	25667	0.619	0.007	57.5	0.079	161
Finland*	338145	5.27	1200*	10000**	100000	0.296	0.004	227.7	0.029	1897
France	643472	66.26	2010	10900	167500	0.260	0.003	30.3	0.017	165
Germany	357022	80.99	2590	16200	215833	0.605	0.007	32.0	0.045	200
Greece	131957	10.82	135	648	11250	0.085	0.001	12.5	0.005	60
Ireland	70273	4.83	266	1240	22167	0.315	0.004	55.1	0.017	256
Italy	301340	61.68	257	1500	21417	0.071	0.001	4.2	0.005	24
The Netherlands	41543	16.89	690	5000	57500	1.384	0.017	40.8	0.120	296
Norway	323802	5.15	1300	8260	108333	0.335	0.004	252.4	0.025	1609
Poland	312685	38.35	390	2000	32500	0.104	0.001	10.1	0.006	52
Spain	505370	47.74	120	463	1000	0.002	0.000	2.51	0.001	10
Sweden	450295	9.72	5600	51920	466670	1.036	0.012	576.1	0.115	5341
Switzerland	41277	8.06	1684	10897	140333	3.400	0.041	208.9	0.264	1352
United Kingdom	24'610	63.74	280	1800	23333	0.096	0.001	4.39	0.007	28

Total: 18.18 GW$_{th}$ 131.67 PJ/yr

Note: Data of installed MW$_{th}$ and of produced TJ/yr for WGC 2015 from the Appendix, country size, and population from https://en.wikipedia.org/wiki/List_of_sovereign_states_and_dependent_territories_in_Europe, retrieved September 13, 2015. Countries with more than 100 MW$_{th}$ installed capacity are listed.
* No Country Report has been submitted from Finland to WGC 2015.
** Rough estimates for 2013.

Table 7.2   Ranking results of the top five countries in the respective categories.

Country	Installed capacity (kW_{th}) per country area (km^2)
Switzerland	41
The Netherlands	17
Sweden	12
Austria	10
Denmark	7

	Installed capacity (W_{th}) per capita
Sweden	576
Norway	252
Finland	228
Switzerland	209
Austria	102

	Annual heat production (GJ/yr) per country area (km^2)
Switzerland	264
The Netherlands	120
Sweden	115
Denmark	79
Austria	59

	Annual heat production (MJ/yr) per capita
Sweden	5341
Finland	1897
Norway	1603
Switzerland	1352
Denmark	610

7.4. Medalists

The achievements of the top five countries in GHP development can be further specified on the basis of the latest status as reported in the WGC 2015 direct use summary. For this, the ranking numbers

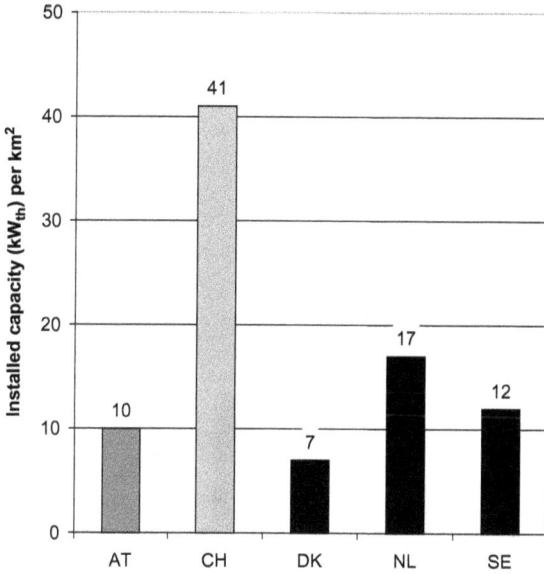

Figure 7.4   Visualization of ranking results, installed capacity per area.

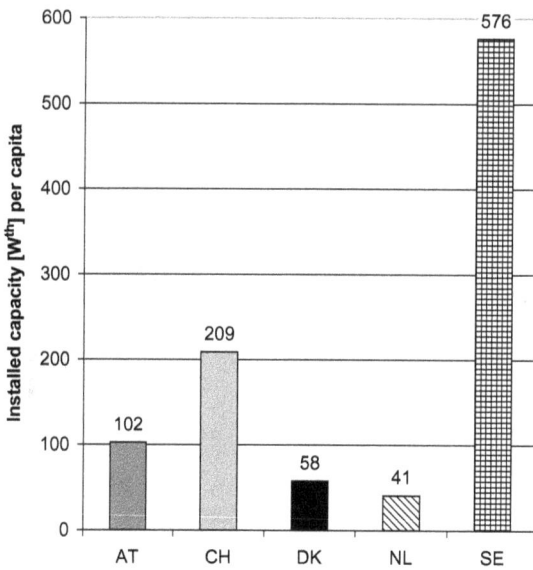

Figure 7.5   Visualization of ranking results, installed capacity per capita.

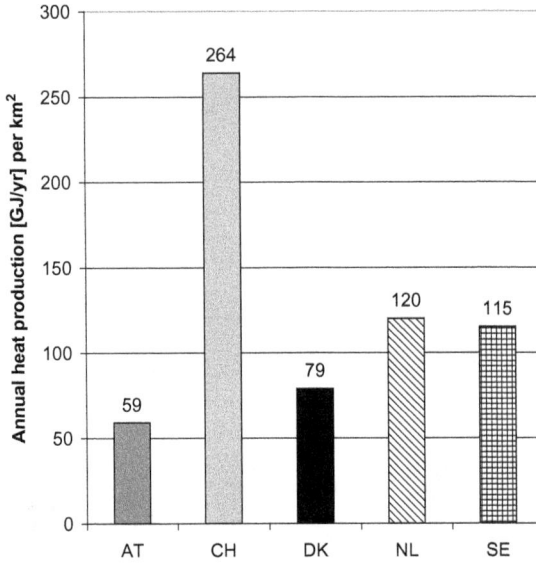

Figure 7.6   Visualization of ranking results, annual heat production per area.

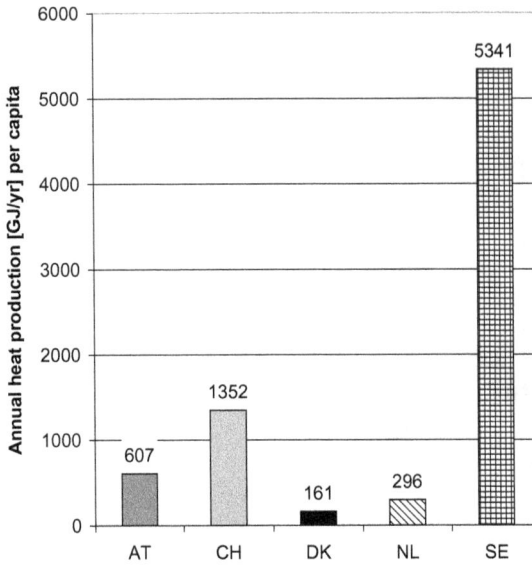

Figure 7.7   Visualization of ranking results, annual heat production.

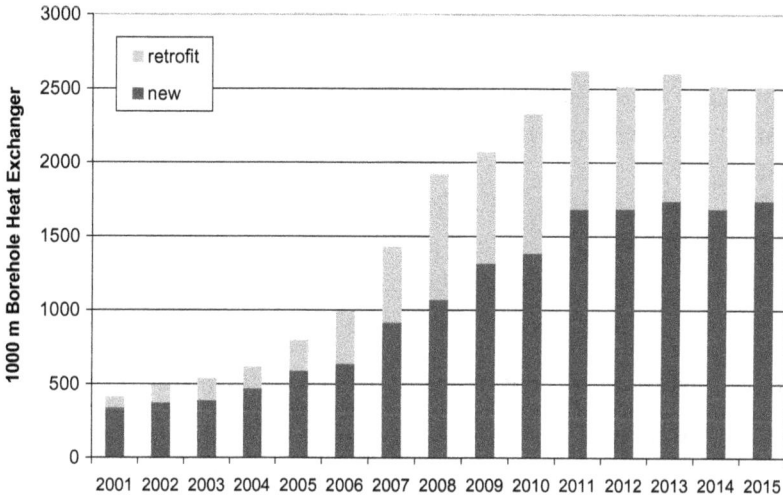

Figure 7.8   Annual drilling activities in Switzerland over the years 2001–2015. The substantial activities in retrofit lead to CO_2 emission reduction.

Source: FWS (2016).

Table 7.3   Top five countries in ranking orders.

Rank	MW_{th}/area	MW_{th} per capita	TJ/yr per area	TJ/yr per capita
1	Switzerland	Sweden	Switzerland	Sweden
2	The Netherlands	Norway	The Netherlands	Finland
3	Sweden	Finland	Sweden	Norway
4	Austria	Switzerland	Denmark	Switzerland
5	Denmark	Austria	Austria	Denmark

Table 7.4   Ranking medalists.

Country	First rank: Gold	Second rank: Silver	Third rank: Bronze
Finland		1	1
Norway		1	1
Sweden	2		2
Switzerland	2		
The Netherlands		2	

of Table 7.2 can be taken. Then from these numbers, the ranking successions can be derived; the results are given in Table 7.3.

Finally, in the sense of the Olympic spirit, medals can be distributed, based on the ranks in Table 7.3. Table 7.4 shows the medal winners.

The ranking medal winners clearly demonstrate that countries smaller than the big ones like Germany, France, and United Kingdom have shown quite some achievements.

7.5. Current and Possible Future Development Trends

First, the top five countries with already >1 GW_{th} installed capacities (as reported in the WGC 2015 Direct Use Summary Paper) should be considered. It is probable that the growth will still continue, but at a slower pace. In Germany and Switzerland, certain saturation is apparent; nevertheless the development still continues at a rather constant but substantially high level. An example of this is given in Figure 7.8: here the yearly drilled meters for borehole heat exchangers in Switzerland are plotted over the years 2001–2014. It is evident that the drilling activities are — after a nearly exponential growth phase in the years 2001–2011 — are now rather constant, but at a very high level of annually around 2500 km.

Sweden will certainly further develop their GHPs, and Norway probably will follow suit; France might even speed up thanks to governmental incentive schemes.

Several other countries are in the 100 to several 100 MW_{th} installed capacity range (as reported in the WGC 2015 Direct Use Summary Paper): Austria, Belgium, Czech Republic, Denmark, Greece, Ireland, Italy, the Netherlands, and the United Kingdom. Some of them show impressive growth over the last few years (especially the Netherlands and Greece) (see Table 7.5).

Several countries show high growth rates. Hopefully, their growth can continue and that other countries will follow their examples.

There is a large number of numerous and various activities and programs, completed or ongoing, of the European Geothermal Energy Council (EGEC) to promote GHP development

Table 7.5 Growth over the past few years in countries of the range 100 to several 100 MW$_{th}$ installed capacity.

Country	MW$_{th}$ as in WGC2010	MW$_{th}$ as in WGC2015	Growth (%)
Austria	600	840	40
The Netherlands	175	690	294
Denmark	160	320	100
Czech Republic	147	300	104
United Kingdom	182	280	54
Ireland	151	266	76
Italy	231	251	11
Belgium	114	199	75
Greece	50	135	170

and dissemination in Europe (GROUNDREACH, GROUNDHIT, GROUNDMED, GEOTRAINET, THERMOMAP, REGEOCITI-TIES, HEAT UNDERYOURFEET) . These are summarized in Part B of this chapter.

PART B: TECHNICAL DEVELOPMENT AND STATUS

7.6. Technical Fundamentals of Geothermal Heat Pumps

In terms of the number of installations, installed capacity and energy produced, shallow geothermal energy is by far the largest sector of geothermal energy use in Europe (Figure 7.9). It enjoys the widest deployment among European countries, with very few countries not having shallow geothermal installations at all.

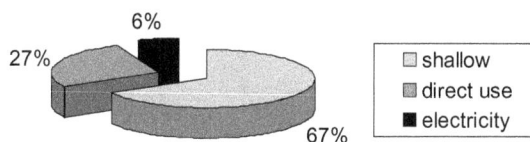

Figure 7.9 Share of installed capacity in the three geothermal subsectors in Europe as of 2015 (after Antics *et al.*, 2016).

Shallow geothermal energy is available everywhere, and it is harnessed typically by ground source heat pump (GSHP) installations, using the heat pump to adjust the temperature of the heat extracted from the ground to the (higher) level needed in the building, or to adjust the temperature of heat coming from building cooling to the (lower) level required to inject it into the ground. The main technologies used to connect the underground heat to the building system (Figure 7.10) comprise

- open-loop systems, with direct use of groundwater through wells;
- closed-loop systems, with heat exchangers of several types in the underground; horizontal loops, borehole heat exchangers (BHE), compact forms of ground heat exchangers, thermoactive structures (pipes in any kind of building elements in contact with the ground), etc.

While Figure 7.10 shows the exterior form of different ground coupling options, Figure 7.11 details the internal constellations possible for closed shallow geothermal systems. They differ in the type of heat carrier medium inside the ground circuit, and in the way this circuit is coupled to the heat pump refrigeration cycle. The most common constellation is the use of a fluid as heat carrier (typically water with the addition of an antifreeze agent), which is circulated through the ground loop by pumping.

Direct expansion systems are characterized by the extension of the refrigeration cycle into the ground loop, i.e., the heat carrier is

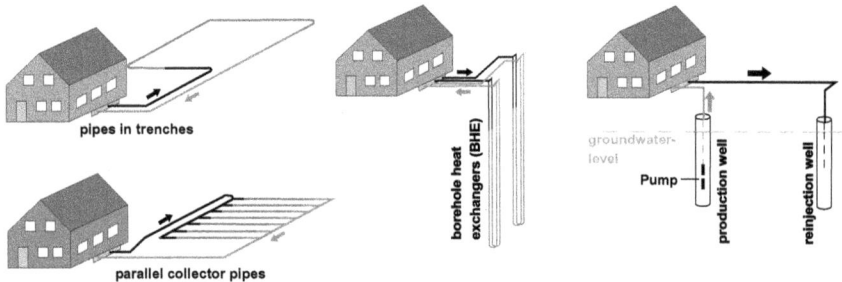

Figure 7.10 Schematic of the most common ground-coupling methods, from left: horizontal loops, (BHE, a.k.a. vertical loops), and groundwater wells.

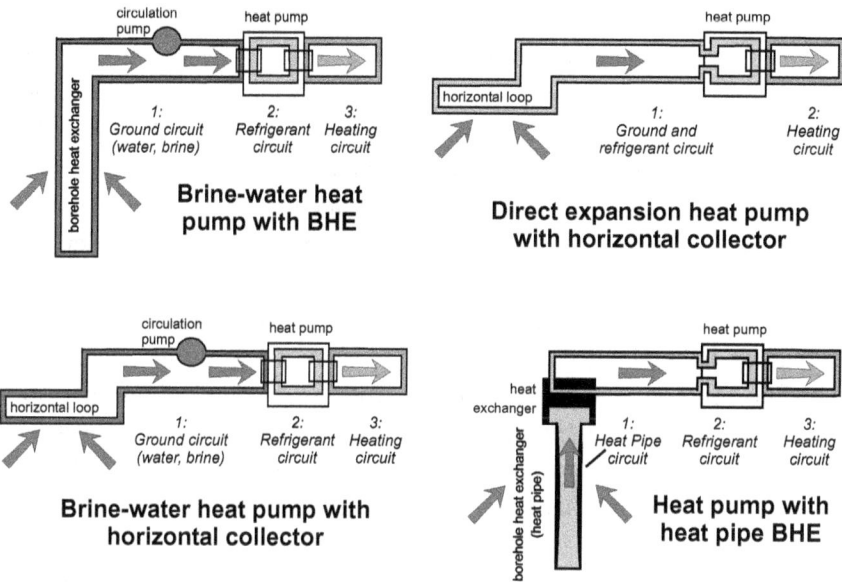

Figure 7.11   Possible ground loop circuits: fluid (brine) circuit for vertical and for horizontal loops (left), direct expansion (DX) circuit for horizontal loop (upper right) and heat pipe circuit for vertical loop (lower right).

the working medium of the heat pump, and a two-phase-flow (liquid/steam) occurs inside the ground loop. In practice, direct expansion (DX) has been applied successfully to GSHP with horizontal loop, while the combination with vertical loops resulted in problems with compressor oil return, etc. Heat pipes make use of a two-phase system inside a single, vertical pipe. The working medium with low boiling point is evaporated by the earth's heat in the lower section of the pipe. The resulting steam rises to the top of the pipe due to its lower density, and transfers the heat to the refrigeration circuit via a heat exchanger. The steam thus cools down and condenses again, flowing back in liquid form on the pipe wall toward the bottom of the pipe. While both the brine systems and the DX systems can be used both for heating and cooling, the heat pipe is suitable for heating purposes only, as no heat can be transported down into the ground (the driving force is provided by gravity, which works only in one direction).

Shallow geothermal installations intended to change the underground temperature periodically (e.g., seasonally) fall under the term underground thermal energy storage (UTES). The delineation between GSHP and UTES is not sharp, and among the larger installations, only a minority are "pure GSHP". Large GSHP plants in most cases have a high share of the annual energy turnover inside the BHE field or the aquifer, and not with the surrounding or underlying round, and thus qualify for the term "storage". In all these large installations, it is crucial to pursue a long-term balance of heat extracted from the ground and injected into the ground.

The different natural ground temperatures throughout Europe, from 2°C to 3°C near the polar circle to about 20°C in the very south of Europe, have a great influence on the options and design for shallow geothermal installations. In combination with the building loads, as such also influenced by the climatic zone the site is in, and the thermal and hydraulic parameters of the underground on site, the plant design has to guarantee that temperatures in the underground systems are kept within a given range in the long term. This temperature range is defined on one side by the technical (thermal) requirements of the building system, and on the other side by environmental considerations concerning the groundwater and ground at the specific site.

Often buildings have a rather unbalanced heating and cooling demand, either given by their climatic surroundings (very cold and warm climates), or by the specific use of the building (there are, for example, shopping malls even in Northern Europe that require virtually no heating, but a lot of cooling). In these cases, hybrid systems are designed to cover as much load as possible from the geothermal system, and to balance the heat in the underground by separate sources like cold air in winter or at nighttime, waste heat, solar heat, etc. Using all the different design options available to geothermal design allows for small and large, energy-efficient, economic, and reliable installations all over Europe. A nice example here is the case of the Swedish company IKEA. A growing number of stores from Sweden (Figure 7.12) to Spain are equipped with shallow geothermal technology of different type, and adapted to the respective geological and climatic situation (Table 7.6).

Figure 7.12   IKEA store in Uppsala, Sweden, under construction in 2009; the BHE field is located under the area in the foreground (Photo: Sanner).

Table 7.6   Examples of IKEA-stores in Europe using shallow geothermal technology.

Country	City	Technology	Size
Austria	Klagenfurt	thermal piles	420 piles
Germany	Lübeck-Dänischburg	BHEs	215, each 150 m deep
Ireland	Dublin	BHEs	Total 2 MW capacity
Italy	Milano-Corsico	BHEs	304, 87-125 m deep
Norway	Oslo-Billingstad	BHEs	86, each 200 m deep
Sweden	Helsingborg-Väla	BHEs	67, each 150 m deep
Sweden	Karlstad	BHEs	100, each 120 m deep
Sweden	Umeå	BHEs	50, each 200 m deep
Sweden	Uppsala	BHEs	100, each 168 m deep
Spain	Jerez de la Frontera	BHEs	50, each 130 m deep
Spain	Madrid-Alcorcón	BHEs	45, each 100 m deep
Spain	Valencia-Alfafar	groundwater wells	3 wells > 100 m deep
Examples from outside Europe			
USA	Centennial, Denver CO	BHEs	130, each 165 m deep
USA	Conshohocken, PA	BHEs	180, each 200 m deep

All material for GSHP systems today is available from man-ufacturers, in proven quality: prefabricated BHE, grouting mate-rial, pipes, manifolds, and heat pumps. Methods for determining the ground parameters (thermal and hydraulic) are available, design rules and calculation methods have been developed, and guidelines and standards set the frame for reliable and durable installations.

7.7. Borehole Heat Exchangers

The most common ground-coupling technology in shallow geother-mal systems is the BHE, also called the vertical loop. The advantages of BHE are the wide scope of possible applications, lack of need for specific hydrogeological conditions, relatively simple installation, and virtually no maintenance. However, for this kind of closed system, the heat transfer from the ground into the fluid inside the BHE pipes and vice versa is driven by temperature difference only. This limits the potential heat exchange capacity and requires appropriate sizing of borehole depth and number. Good knowledge of the ground ther-mal parameters is crucial for design calculations; the mobile thermal response test (TRT) offers a proven tool to determine the values on site (Figure 7.13).

Because of the need to circulate a fluid down into the earth and up again, there are only few basic design options for BHE:

- Coaxial (or concentric) pipes, also called pipe-in-pipe.
- U-pipes (two or more simple pipes connected at the bottom).
- Only for heat pipes, a single pipe is sufficient, as the vapor can rise upward in the center of the pipe while the condensate flows down alongside the pipe walls.

Over the course of more than 60 years of BHE development, var-ious design alternatives have been developed and tested. Due to cost effectiveness, only a few rather simple designs prevail (Figure 7.14). These BHE are inserted in boreholes, and the remaining annulus between the pipes and the borehole wall is either filled with a special grouting material, or with water if the borehole wall is stable (the latter limited to Scandinavia).

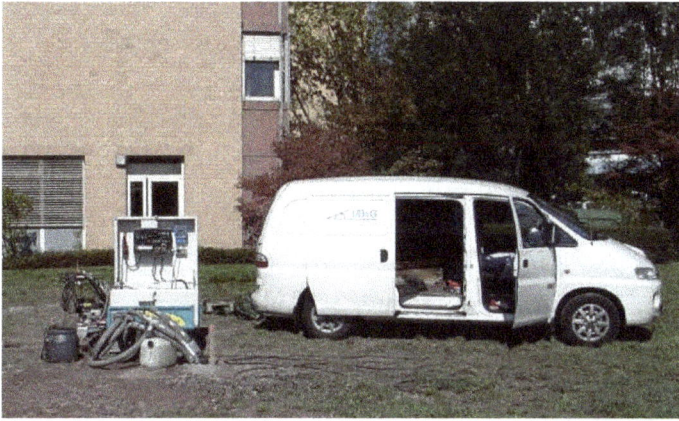

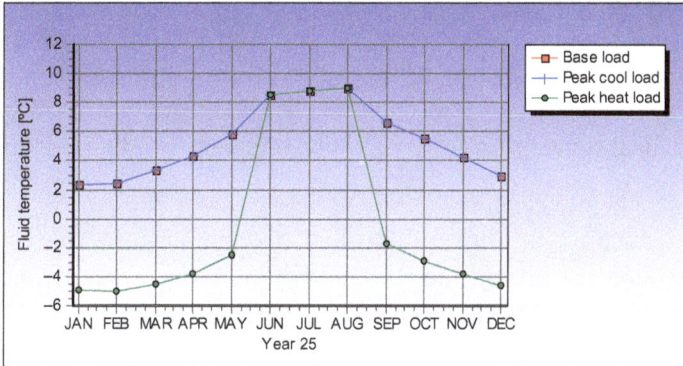

Figure 7.13 Example of TRT for determining ground parameters (top, photo UBeG) and calculation of BHE layout using EED (bottom).

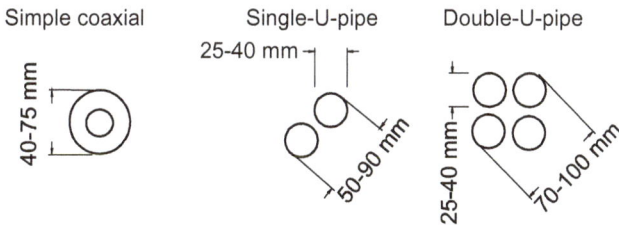

Figure 7.14 Cross-sections of three most frequent BHE types.

A useful tool for comparing different installations of BHE is the specific heat extraction rate. This is the maximum thermal capacity at the heat pump evaporator (refrigeration capacity), divided by the total length of BHE, given in watt per meter BHE length (W/m). In the early times of BHE in Europe around 1980, a value of 50 W/m was given as a standard value for Germany, and 55 W/m for Switzerland. These values were used for design of residential GSHP at that time — and 50 W/m is still used as a crude rule of thumb for many smaller installations until today. However, the actual specific heat extraction possible in a certain project depends strongly upon ground conditions (thermal conductivity), system requirements (operating hours), system size (number and distance of BHE, interference), etc. (Sanner, 1999). So a BHE system never should be designed following a rule of 50 W/m of heat extraction, and the specific heat extraction value only be used for comparison after a thorough design calculation has been made.

The heat transport in a BHE system can be divided into two stages:

- The transport in the undisturbed ground around the borehole (controlled mainly by the thermal conductivity of the ground, λ).
- The transport from the borehole wall into the fluid inside the pipes (or vice versa), controlled by the type of grouting, the pipe material, the borehole and pipe geometry, etc., and given as a summary parameter r_b (borehole thermal resistance).

For BHE design, only the parameters controlling r_b can be influenced by engineering, as the ground outside the borehole cannot be changed. The specific heat extraction rate of a BHE can only be calculated for a certain installation, taking into account all the parameters mentioned above. The best BHE would be a system with $r_b = 0$ K(W/m), i.e., a spontaneous heat transfer between borehole wall and fluid. This can be achieved only theoretically, but can act as a benchmark for determining the efficiency of an actual BHE system. This efficiency is called Hellström efficiency and is given as

$$\eta_H = \frac{\text{Sustainable heat extraction possible in a given project}}{\text{Heat extraction with } r_b = 0}$$

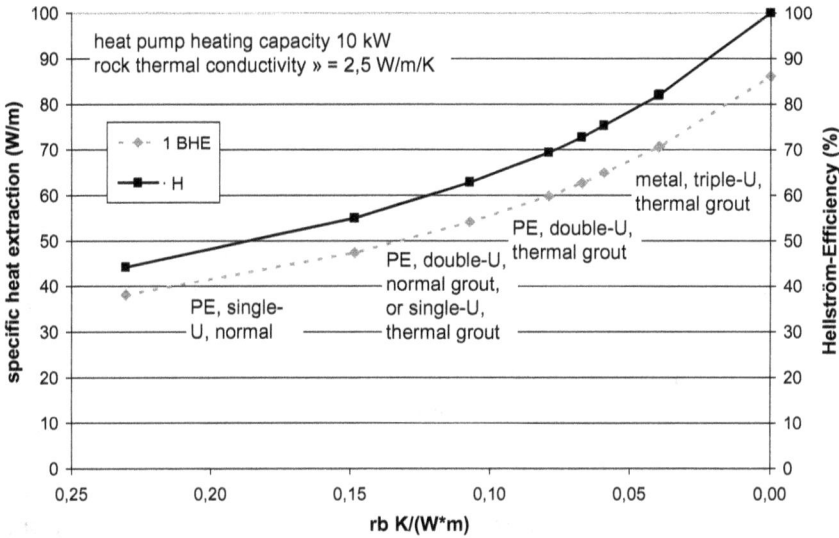

Figure 7.15 Example for Hellström efficiency of different BHE types in a single-family house under average ground conditions; the maximum heat extraction rate at $\eta_H = 100$ cannot exceed ca. 85 W/m for this specific ground type and building.

A value of $\eta_H = 100$ stands for the theoretical maximum. The Hellström efficiency always relates to a given building load and ground thermal parameters; the maximum efficiency of a BHE under given load circumstances (i.e., permissible temperature difference ground fluid and planned operation time) can be calculated and plotted against r_b (Figure 7.15). The quality of a BHE type thus can be benchmarked using r_b and the Hellström efficiency, η_H. It depends mainly on pipe material, pipe size, pipe configuration, and filling of the annulus.

7.8. Factors Influencing the Market Development of Ground Source Heat Pumps

The market for GSHP today is in difficulty nearly anywhere. While in some mature markets the situation still is rather stable, in others a decrease can be seen. In parts of Germany, this can be attributed to continuously stricter regulation, causing delays and higher cost (see below). What are the main reasons for the current low in the market?

(a) Not enough awareness about this technology and its advantages. Especially architects, the building sector, local authorities need to be informed better.

(b) Cost intensity is an issue, in particular for the investment part. Because of the drilling, GHPs can be considered as a capital-intensive technology in comparison with other small scale applications.

(c) Quite unfavorable competition with gas: Geothermal heat technologies are heading for competitiveness, but support is still needed in certain cases, notably in emerging markets and where a level playing field does not exist. In addition, there is a need for an in-depth analysis of the heat sector, including about the best practices to promote geothermal heat, the synergies between energy efficiency and renewable heating and cooling, and barriers to competitiveness. As GHPs can be considered a mature and competitive technology, a level-playing field with the fossil fuel heating systems would eventually allow phasing out subsidies for shallow geothermal in the heating sector.

(d) Regulations can be a barrier, either through over-regulation, or with unclear procedures. Even a lack of regulation can turn into a barrier over time, with drilling done in unsuitable places or with insufficient care. We need simplifications in some countries, more stable regulations in other, and a better knowledge within the authorities almost everywhere.

(e) Bad publicity with a few projects causing problems in Germany and France, receiving high press coverage.

An example of regulation acting as a barrier to ground source heat pump installation can be seen in Germany. The decrease in annual GSHP installation is quite different in some of the German states (Länder). Numbers on licenses granted annually for BHE drilling and installation are available in two adjacent states, Hessen and Rheinland-Pfalz (Figure 7.16). These data do not reflect exactly the number of BHE installed, as one installation might comprise more

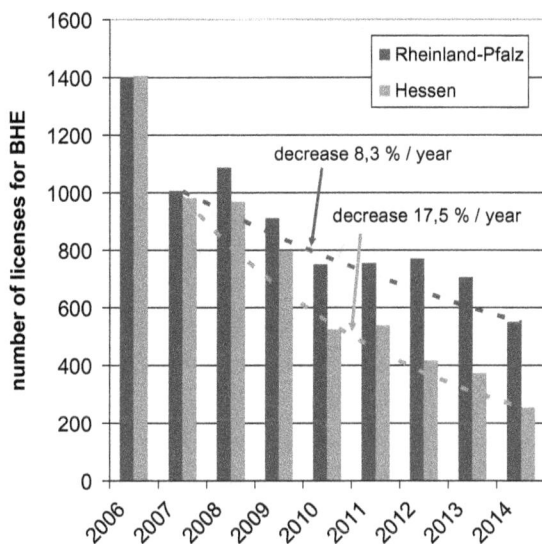

Figure 7.16 Annual number of licenses granted for BHE in the German states Hessen and Rheinland-Pfalz, after data from the relevant state authorities (average decrease calculated by the author).

than just a single BHE, and also not all projects where a license was obtained might have gone ahead. Geology, climate, and energy cost in the two states is comparable, while from 2007 to 2014 (to exclude the extraordinary year 2006), the average decrease of licenses in Hessen is more than double as fast as in Rheinland-Pfalz. The decrease in the number of GSHP sales for the whole of Germany is again much slower, with an average of just 5% for the same time horizon, 2007–2014. While this value is not directly comparable to the number of BHE licenses, as other ground coupling methods than BHE are included, and some plants might have more than one heat pump installed, the trend, however, can be seen clearly, with the market for BHE in Hessen shrinking much faster than in the rest of Germany. A similar development can be expected for Baden-Württemberg, where extremely rigid regulations have been put in place, but exact numbers are not public yet.

Ensuring good quality of design, components, material, and installation is key for a favorable long-term market development.

Here the issues of standards and training play an important role. Standards for GSHP installations date back to the 1990s, with a German standard related to protection of ground and groundwater (DIN 8901), the Swiss guidelines on ground-coupling technologies (AWP T1–T5, no longer valid today), and the Swedish guideline "Normbrunn" (on wells for water and for energy). In 1998, the draft of German guideline VDI 4640 offered the first comprehensive standard for the shallow geothermal sector. Today, a number of standards exist in different European countries, and the setting of European standards through CEN is envisaged. Table 7.7 provides an overview of the current status of standards for ground source heat pumps.

Training for installers of shallow geothermal systems is provided by professional associations or industrial associations, often in cooperation with the existing educational institutions, in countries with a mature GSHP market; this includes, for example, Germany, Netherlands, Sweden, and Switzerland. In order to help other countries to set up training schemes and for all countries to have training based upon certain minimum standards, the program Geotrainet was developed. It started as a project from 2008 to 2011 with support by the European Union (EU) within the IEE program. After lengthy preparations, the new association named Geotrainet was founded in November 2013 in Lund, Sweden, and is registered in Brussels since the spring of 2014. This association is destined to continue the coordination of training activities, to set the curricula, to provide common training material, etc.; the members are national training coordinators in the different states (mostly national associations). An European Training Committee was set up in 2015 to decide on the curricula, minimum standards, and training material.

The ReGeoCities project (2012–2015) worked on regulatory issues on shallow geothermal, involving several stakeholders. ReGeoCities project results indicate a widespread lack of awareness among public authorities and the general public regarding ground source heat pumps. For this reason, it was deemed important — as part of the dissemination and communication activities — to launch a communication campaign under the name The Heat Under

Table 7.7   Existing standards and guidelines on GHPs in Europe.

Country	Code	Title (in English translation where appropriate)	Year
Europe	EN 15450	Heating systems in buildings: Design of heat pump heating systems	2007
Europe	EN ISO 17628	Geotechnical investigation and testing: Geothermal testing — Determination of thermal conductivity of soil and rock using a BHE	2015
Austria	ÖWAV-Regelblatt 207	Thermal use of groundwater and underground: Heating and cooling	2009
Germany	DIN 8901	Refrigerating systems and heat pumps: Protection of soil, ground and surface water	2002
Germany	VDI 4640-1	Thermal use of the underground: Fundamentals, approvals, environmental aspects	2010
Germany	VDI 4640-2	Thermal use of the underground: GSHP systems	2001 (2015 draft)
Germany	VDI 4640-3	Thermal use of the underground: UTES	2001
Germany	VDI 4640-4	Thermal use of the underground: Direct uses	2004
Germany	VDI 4640-5	Thermal use of the underground: Thermal response test	2016 (draft)
France	NF X10-960-1	Boreholes for water and geothermal: vertical BHEs, General issues	2013
France	NF X10-960-2	Boreholes for water and geothermal: vertical BHEs, pipe loops of polyethylene 100 (PE 100)	2013

(*Continued*)

Table 7.7   (*Continued*)

Country	Code	Title (in English translation where appropriate)	Year
France	NF X10-960-3	Boreholes for water and geothermal: vertical BHEs, pipe loops of cross-linked polyethylene (PE-X)	2013
France	NF X10-960-4	Boreholes for water and geothermal: vertical BHEs, pipe loops of polyethylene with higher temperature resistance (PE-RT)	2013
France	NF X10-970	Boreholes for water and geothermal: vertical BHEs, Installation, commissioning, maintenance, abandonment	2011
Italy	UNI 11466	Geothermal systems with heat pump: requirements for the dimensioning and design	2012
Italy	UNI 11467	Geothermal systems with heat pump: requirements for installation	2012
Italy	UNI 11468	Geothermal systems with heat pump: environmental requirements	2012
Italy	UNI/TS 11487	Geothermal systems with heat pump: requirements for the installation of DX systems	2013
Italy	UNI 11517	Geothermal systems with heat pump: requirements for the qualification of companies installing geothermal heat exchangers	2013
Spain	UNE 100715-1	Design, installation and maintenance of shallow geothermal installations: closed vertical systems	2014
Sweden	SGU Normbrunn-07	Drilling of wells for energy and water	2008
Switzerland	SN 546 384/6	Borehole heat exchangers	2010

(*Continued*)

Table 7.7 (*Continued*)

Country	Code	Title (in English translation where appropriate)	Year
Switzerland	SN 546 384/7	Use of the heat of the groundwater	2015
UK	DECC MIS 3005	Requirements for contractors undertaking the supply, design, installation, set to work commissioning, and handover of microgeneration heat pump systems	2011
UK	GSHPA	Closed-loop vertical borehole: Design, installation, and materials standards	2011
UK	GSHPA	Thermal Pile: Design, installation, and materials standards	2012

Your Feet, which targets mainly public authorities, architects, and builders, but also the general public. The campaign resources can be found online at www.heatunderyourfeet.eu, and the campaign can be contacted through twitter@heatunderurfeet.

The campaign uses a narrative that aims at making the technology more accessible and easy to understand for the target groups at which the campaign is aimed. This is done partly through the use of graphics, which have the advantage of making highly technical content more accessible. The key messages focus on illustrating the advantages of GHPs in the heating and cooling sector: The heating and cooling sector for buildings is today, for the large majority, dominated by the use of fossil fuels such as natural gas and heating oil. This means it is contributing heavily to costly fossil fuels imports, exposure to price volatility and security of supply, and production of harmful greenhouse gas (GHG) emissions.

Geothermal heat pumps offer the perfect solution to replace fossil fuels and reverse the unsustainable situation. Their wide range of application, their efficiency, their reliability, all strongly contribute to provide affordable heat, to reduce emissions, and to save primary energy.

7.9. Technological Development Perspectives

The possible technological development for ground source heat pumps was investigated within the Strategic Research Priorities for Geothermal Technology (RHC Platform, 2012) of the European Technology and Innovation Platform for Renewable Heating and Cooling (RHC-ETIP), and the most promising areas were identified. Those promising topics then were made part of both the Geothermal Roadmap (RHC Platform, 2014a) and the Common Roadmap (RHC Platform, 2014a), and eventually became part of the work-relevant program within the EU support framework for R&D, "Horizon 2020". The main key performance indicators (KPI) and objectives for GSHP as to the Geothermal Roadmap are as follows:

- The performance of GHP systems improved substantially since their introduction in Europe in the 1970s. The first plants were installed in Sweden, Germany, and Switzerland, and used for heating only. In these regions, the typical efficiency, expressed as Seasonal Performance Factor, increased from below 3 in the 1980s to well above 4 today, and with continued R&D, average values in the order of 5 seem feasible for 2020.
- Component efficiency improvement: The most popular ground-coupling technology is the BHE; a good efficiency of a BHE results in a small temperature loss between the ground and the fluid inside the BHE. This temperature loss is controlled by the borehole thermal resistance, r_b. This performance indicator has been reduced by more than 40% over the last 10 years. The overall impact of this value to a defined shallow geothermal system is given by the Hellström efficiency, which has increased from below 60% to about 75% in state-of-the-art installations over the past 10 years. There is still room for improvement, so provided the technology progress is continued, efficiencies of about 80% in 2020 seem achievable.
- The cost shows a steady reduction in the last decades. A study of the Swiss Heat Pump Association (Fördergemeinschaft

Wärmepumpen Schweiz, FWS) calculated the cost for a BHE system (drilling, heat exchanger, and heat pump) for a small house, and found a reduction of 27.5% over 12 years, from 1992 to 2004. While the initial cost of a BHE system has decreased slightly, improvements in efficiency, which result in less energy being used to operate the system, have led to a substantial cost reduction overall.

And the "Common Roadmap" summarizes the KPIs as follows:

- A seasonal performance factor in the order of 5 for 2020.
- A Hellström efficiency (a measure of the impact of borehole thermal resistance) of about 80% in 2020.
- A further decrease in energy input and reduced costs for operating the GHP system.

Several EU projects meanwhile have started to help in achieving the objectives, and more can be expected in the future within Horizon 2020. Two projects address the topic "Improved vertical borehole drilling technologies to enhance safety and reduce cost of BHE installations — Improved installation technologies and geometries for ground heat exchange technology" and also the topic "Improved pipe materials for BHE and horizontal ground loops — New pipes for higher temperatures — Better thermal transfer fluid" meanwhile has been opened in the work program.

A cooperative action was started to improve the understanding of "geoactive structures" (thermal piles, thermal foundations, thermal diaphragm walls, etc.). Relevant national research centers throughout Europe are jointly addressing the topic "European-wide Geoactive Structures Alliance — Development of a network of laboratories to create four testing sites."

Activities are also ongoing in the field of geological information as a basis for design of GSHPs. National geological surveys have started to provide relevant information in online GIS formats, for example, in France, in some German states, in the Flemish region of Belgium, and elsewhere. These activities answer the topic "Creation of a new European wide database to map conductivities and potential (to 100 m depth) and feasibility of vertical BHE systems" in the roadmap. The newly launched European Geological Data Infrastructure (EGDI), coordinated by EuroGeoSurveys, the association of the relevant national institutions, could help in harmonizing the regional and national GIS datasets and eventually provide a Europe-wide tool.

In the roadmaps, and subsequently in the Horizon 2020 work program, nontechnical issues are also addressed. These are of high importance to overcome barriers in wider application of ground source heat pumps, as explained earlier in this chapter. The respective topic reads: "Non-technical provisions: measures to increase awareness, harmonization of shallow geo-standards, shallow geothermal installer EU wide training certificate, shallow geothermal Smart City deployment policy along the line of previous projects."

With all the R&D efforts supported by EU, national and regional authorities, or executed and financed directly by the industry, further improvements in efficiency, cost reduction (both in installation and operation), and environmental soundness can be expected, promising a sustainable future for GHP systems.

References

Antics, M., Bertani, R., and Sanner, B. 2016. Summary of EGC 2016 Country Update Reports on Geothermal Energy in Europe, Proc. EGC 2016, paper keynote-1, 16 p., Strasbourg.

EurObservER 2013. Heat Pumps Barometer, 18 p., Observ'ER, Paris.

EurObservER 2015. Heat Pumps Barometer, 12 p., Observ'ER, Paris.

FWS 2016. Statistik Erdwärmesonden Bohrmeter 2015. From http://www.fws. ch/statistiken.html#statistik-2015, retrieved October 5. Fachvereinigung Wärmepumpen Schweiz, Bern, Switzerland.

Freeston, D.H. 1995. Direct Uses Of Geothermal Energy 1995 (Preliminary Review). Proceedings World Geothermal Congress 1995, Florence, Italy, pp. 15–25.

Lund, J., Sanner, B., Rybach, L., Curtis, R., and Hellström, G. 2003. Ground source heat pumps — A world review. *Renewable Energy World*, July–August 2003, pp. 218–227.

Lund, J.W. and Freeston, D.H. 2000. World-wide Direct Uses of Geothermal Energy 2000. Proceedings World Geothermal Congress 2000, Kyushu, Tohuku, Japan, 21p.

Lund, J.W., Freeston, D.H., and Boyd, T.L. 2005. World-wide Direct Uses of Geothermal Energy 2005. Proceedings World Geothermal Congress 2005, Antalya, Turkey, 20 p.

Lund, J.W., Freeston, D.H., and Boyd, T.L. 2010. Direct Utilization of Geothermal Enwrgy 2010: Worldwide Review. Proceedings World Geothermal Congress 2010, Bali, Indonesia, 23 p.

Lund, J.W. and Boyd, T.L. 2015. Direct Utilization of Geothermal Energy 2015: Worldwide Review. Proceedings World Geothermal Congress 2015, Melbourne, Australia, 31 p.

RHC Platfom 2012. Strategic Research Priorities for Geothermal Technology, 70 p., Brussels (www.rhc-platform.org).

RHC Platfom 2014a. Geothermal Technology Roadmap, 36 p., Brussels (www.rhc-platform.org).

RHC Platfom 2014b. Common Implementation Roadmap for Renewable Heating and Cooling Technologies, 62 p., Brussels (www.rhc-platform.org).

Sanner, B. 1999. Kann man Erdwärmesonden mit Hilfe von spezifischen Entzugsleistungen auslegen? Geothermische Energie 26–27/99, pp. 1–4, Geeste.

Appendix: Data Base of European GHP Development Trends 2000–2015

Remarks in italics

Country areas (km^2) and population (millions) from https://en.wikipedia.org/wiki/List_of_sovereign_states_and_depend ent_territories_in_Europe (retrieved September 13, 2015).

Albania 28,748 km^2, 3.02M
2000 No data; 2005 No data; 2010: Geothermal heat pumps use is 1.9 MW$_{th}$ and 31.9 TJ/yr.
2015: GHPS with 1.2 MW$_{th}$ and 11.5 TJ/yr for GHPs. *Less than in 2010*.

Armenia 29,743 km^2, 3.06M
2000 No data; 2005: estimated installed capacity of 0.5 MW$_{th}$ and annual energy use of 3.3 TJ.
2010 No data; 2015 No data.

Austria 83,871km^2, 8.22M
2000: There is an estimate 228 MW$_{th}$ installed capacity and annual production 1094 TJ.
2005: An assumed 25,000 GHPs are installed, estimated to provide 300 MW$_{th}$ of capacity and an annual production of 1,450 TJ. 2010: Geothermal heat pumps: capacity of 600 MW$_{th}$ and annual production 2880 TJ/yr.
2015: GHP capacity of 840 MW$_{th}$ and 4,990 TJ/yr.

Belarus 207,600 km^2, 9.60M
2000: No data; 2005 No data; 2010: 15 large HP and ca. 50 GHPs.
2015: The estimated installed capacity is 6.3 MW$_{th}$ the annual energy use 113.5 TJ/yr.

Belgium 30,528 km^2, 11.24M
2000: No data; 2005: Geothermal heat pump estimates are approximately 60 MW$_{th}$ capacity and 324 TJ annual production.
2010: The estimated GHP installed capacity is 114.0 MW$_{th}$ and the annual production 439.8 TJ.

2015: 198.7 MW_{th} and annual production 756.4 TJ/yr.

Bosnia and Herzegovina 51,197 km², 3.87M

2000 No data; 2005 No data; 2010: Geothermal heat pumps: 0.156 MW_{th} and 1.15 TJ/yr.
2015: Geothermal heat pumps 1.20 MW_{th} and 2.70 TJ/yr.

Bulgaria 110,879 km², 6.92M

2000: An installed capacity is 13.3 MW_{th} is due to GHPs, the annual production is 162.0 TJ. 2005: Geothermal heat pumps 0.3 MW_{th} and 4.4 TJ/yr. 2010: Geothermal heat pumps: 20.63 MW_{th} and 286.23 TJ/yr.
2015: An estimated 10.0 MW_{th} and 47.30 TJ/yr for GHPs.
Strange changes over time.

Croatia 56,594 km², 4.47M

2000 No data; 2005 No data; 2010 No data.
2015: 4.50 MW_{th} and 42.50 TJ/yr for GHPs.

Czech Republic 78,867 km², 10.54 M

2000 No data; 2005: Estimates give 200 MW_{th} for the heat pumps with annual production of 1,130 TJ. 2010: Reported GHPs capacity: 147.0 MW_{th}, estimated annual production 832 TJ (corrected; Lund *et al.*, 2010).
2015: Geothermal heat pumps are estimated at MW_{th}, the annual production at 1700 TJ/yr.

Denmark 43,094 km², 5.57M

2000 No data; 2005: A total of 250 groundwater based heat pumps and 43,000 others are in operation (about 10–20% vertical closed-loop). They are extracting approximately 3900 TJ/yr with an installed capacity estimated at 800 MW_{th}.

2005 data obviously wrong. 2010: Small GHPs: 160 MW_{th} and 1,700 TJ/yr.
2015: GHP installed capacity of 320 MW_{th} and annual production 3,400 TJ/yr.

Estonia 45,228 km^2, 1.26M

2000 No data; 2005 No data; 2010: For 2008, with reported GHP installed capacity of 63.0 MW$_{th}$, the estimated annual production is 356 TJ.

2015: estimated GHP capacity 63.0 MW$_{th}$, annual production 356 TJ. *Same as for 2010.*

Finland 338,145 km^2, 5.27M

2000: The installed capacity is 80.5 MW$_{th}$ and the annual production 484 TJ.

2005: the total installed capacity is 162.5 MW$_{th}$, the annual production 1,220 TJ. 2010: installed GHP capacity 63.0 MW$_{th}$, estimated use 356 TJ/yr. **Less than in 2000.**

2015: *No report has been submitted to WGC 2015. Lund and Boyd (2015) estimated 1,560 MW$_{th}$ installed GHP capacity and annual production around 18,000 TJ, for 2015(!). To comply with the other data in Lund and Boyd (2015), which generally refer to 2013 data, the estimates of 1200 MW$_{th}$ and 10'000 TJ/yr are taken.*

France 643,427 km^2, 66.26M

2000: estimation for 1999 gives an installed capacity of 48 MW$_{th}$ and annual production of 255 TJ.

2005: A little over one million GHP units contribute 16.1 MW$_{th}$ (?) and 468.8 TJ/yr of heat energy.

2010: GHPs 1000 MW$_{th}$ and 7,500 TJ/yr.

2015: 2,010 MW$_{th}$ and 10,900 TJ/yr for GHPs.

Georgia 69,700 km^2, 4.94M

2000 No data; 2005 No data; 2010 No data.

2015: 0.03 MW$_{th}$ and 0.16 TJ/yr for GHPs.

Germany 357,022 km^2, 80.99M

2000: Earth-coupled heat pumps and groundwater heat pumps are estimated to contribute 344 MW$_{th}$ from at least 18,000 installations. The estimated annual production is 1149 TJ. 2005: There are about 30,000 GHPs installed in Germany, rated at a total of 400 MW$_{th}$ and providing 2,200 TJ/yr.

2010: 2,230 MW$_{th}$ and 10,368 TJ/yr for GHPs.
2015: 2,590 MW$_{th}$ and 16,200 TJ/yr for GHPs

Greece 131,957 km^2, 10.82M

2000: For heat pumps the installed capacity is 0. MW$_{th}$ and the annual production 3.1 TJ.
2005: Earth-coupled and groundwater (or seawater) heat pumps have 19 large capacity units totaling 1.0 MW$_{th}$ and producing 5.8 TJ/yr of thermal energy, with other smaller units combining for a grand total of 4.0 MW$_{th}$ and 39.1 TJ/yr.
2010: 50.0 MW$_{th}$ and 270 TJ/yr for GHPs.
2015: 135 MW$_{th}$ and 648 TJ/yr for GHPs.

Hungary 93,028 km^2, 9.92M

2000: Geothermal heat pumps represent 3.8 MW$_{th}$ installed capacity, which is estimated to provide 20.2 TJ annual production.
2005: The estimated capacity for heat pumps is 4.0 MW$_{th}$, with annual production of 22.6 TJ.
2010: 40 MW$_{th}$ and 518 TJ/yr for GHPs.
2015: 42 MW$_{th}$ and 695 TJ/yr for GHPs.

Stagnant.

Iceland 13,000 km^2, 0.317M

2000 No data; 2005: Geothermal heat pumps 4 MW$_{th}$ and 20 TJ/yr. *2 big, not ground-coupled units at Akureyri.*
2010: 4 MW$_{th}$ and 20 TJ/yr for GHPs. *Same as in 2005.*
2015: GHPs: 5 MW$_{th}$ and 17 TJ/yr.

Ireland 70,273 km^2, 4.83M

2000 No data; 2005: GHPs 19.6 MW$_{th}$ and a peak use of 83.6 TJ/yr.
2010: 151.43 MW$_{th}$ and 756.11 TJ/yr are used for GHPs.
2015: GHPs 265.54 MW$_{th}$ and annual production 1,240.54 TJ.

Italy 301,340 km^2, 61.68M

2000: 100 individual heat pumps are estimated to account for 1.2 MW$_{th}$ installed capacity and 6.4 TJ annual production.

2005: GHPs: 120 MW$_{th}$ and 500 TJ/yr.
2010: 231 MW$_{th}$ and 961 TJ/yr for GHPs.
2015: GHPs: 257 MW$_{th}$ and 1500 TJ/yr

Latvia 64,589 km², 2.07M

2000 No data; 2005 No data; 2010: we will use the 2005 report which states an installed capacity of 0.321 MW$_{th}$ and 2.22 TJ/yr. *Strange statement since there are no 2005 data.*
2015: 0.321 MW$_{th}$ and 2.22 TJ/yr for GHPs.

Lithuania 65,300 km², 2.94M

2000 No data; 2005: absorption heat pumps extract energy from 38°C water producing 215,000 MWh of energy from 41 MW$_{th}$ of installed capacity.
2010: Small GHPs provide 34.5 MW$_{th}$ and 305.72 TJ/yr.
2015: Small-scale heat pumps with 76.6 MW$_{th}$ and annual energy use of 678.8 TJ.

Macedonia 25,713 km², 2.09M

2000: No data; 2005 No data; 2010 No data
2015: GHPs: 2.50 MW$_{th}$ and 15.0 TJ/yr.

Netherlands 41,543 km², 16.89M

2000: in 1997 it was estimated that 900 heat pumps were installed. This estimate gives an installed capacity of 10.8 MW$_{th}$ and annual production of 57.4 TJ.
2005: There are a reported 1,600 GHP units in place with an installed capacity of 253.5 MW$_{th}$ and annual production of 685 TJ.
2010: The estimated capacity and use of GHPs in the country is 175 MW$_{th}$ and 1,012.6 TJ/yr for the smaller units (average 7 kW$_{th}$) and 1,219.3 MW$_{th}$ and 9,407.2 TJ/yr for the larger units.
2015: 690 MW$_{th}$ and 5,000 TJ/yr for GHPs.

Norway 323,802 km², 5.15M

2000: an estimate of the number of GHP is made in 1999: this gives an estimated 500 units with 6.0 MW$_{th}$ capacity and annual production 31.9 TJ.

2005: the estimated number of installed units is 13,000 with a capacity of 450 MW$_{th}$, the annual production 2,314 TJ.

2010: the total installed capacity of GSHP is 3,300 MW$_{th}$ producing 25,200 TJ/yr. *Questionably high numbers for 2005.*

2015: The total installed capacity for GHPs is 1,300 MW$_{th}$ and the annual energy use 8,260 TJ.

Poland 312,685 km², 38.35M

2000: in 1999 over 4000 ground-source and groundwater heat pumps were on line, with an installed capacity of 26.2 MW$_{th}$, and the annual production 108.3 TJ.

2010: GHPs: 203.10 MWth and 1,044.5 TJ/yr.

2015: At least 390 MW$_{th}$ and 2,000 TJ/yr for GHPs.

Portugal 92,090 km², 10.43M

2000 No data; 2005: A single heat pump installation (0.2 MW$_{th}$ and 0 TJ/yr); 2010: GHP estimates: 0.3 MW$_{th}$ and 1.1 TJ/yr.

2015: An estimated 15.0 MW$_{th}$ and 90 TJ/yr for GHP.

Romania 238,391 km², 21.70M

2000 No data; 200 No data; 2010: An estimated 5.5 MW$_{th}$ and 29.70 TJ/yr for GHPs.

2015: Estimated installed GHP capacity is about 40 MW$_{th}$ and the annual production is 480 TJ.

Russia 17,098,242 km², 146.27M

2000 No data; 2005: GHPs were reported, consisting of 100 units with an installed capacity of 1.2 MW$_{th}$ and producing 11.5 TJ/y.

2010: 1.2 MW$_{th}$ and 11.5 TJ/yr for GHPs.

2015: 1.2 MW$_{th}$ and 11.5 TJ/yr for GHPs.

Serbia 88,361 km², 7.21M

2000: The installed capacity is 6.0 MW$_{th}$ is for heat pumps, and the annual production 40 TJ.

2005: For GHPs an installed capacity of 6.0 MW$_{th}$ and annual production of 40 TJ is reported.

2010: 9.9 MW$_{th}$ and 83 TJ/yr is reported for GHPs.
2015: 11.0 MW$_{th}$ and 88 TJ/yr for GHPs.

Slovakia 49,035 km^2, 5.44M

2000: the installed capacity for heat pumps is 1.4 MW$_{th}$, and the annual production 12.1 TJ.
2005: Eight GHP installations are reported with a total installed capacity of 1.4 MW$_{th}$ and annual production of 12.1 TJ.
2010: 1.6 MW$_{th}$ and 13.5 TJ/yr for GHPs.
2015: 1.6 MW$_{th}$ and 13.5 TJ/yr for GHPs.

Slovenia 20,273 km^2, 1.99M

2000: No data; 2005: 203 installed GHP units are reported, providing 2.3 MW$_{th}$ of installed capacity and producing 52.8 TJ/yr.
2010: 49.71 MW$_{th}$ and 379 TJ/yr for GHPs.
2015: 85.64 MW$_{th}$ and 501.3 TJ/yr for GHPs.

Spain 505,370 km^2, 47.74M

2000: The installed capacity is 2.6 MW$_{th}$ for heat pumps, and the annual production 46.8 TJ.
2005: No data (no report).
2010: 120 MW$_{th}$ is estimated for GHPs; the annual production is approximately 462.92 TJ. *Too high in comparison to the 2015 data.*
2015: 43.087 MW$_{th}$ and 121.67 TJ/yr for GHPs.

Sweden 450,295 km^2, 9.72M

2000: An estimate of the number of individual heat pumps in operation is reported at about 55,000 with a thermal capacity of 330 MW. The estimated annual use of the 55,000 units is 3,168 TJ.
2005: The installed capacity and the annual production is now estimated at 3,840 MW$_{th}$ and 36,000 TJ. 2010: 4,460 MW$_{th}$ and 45,301 TJ/yr, all as GHPs.
2015: The total GHP capacity is 5,600 MW$_{th}$, the annual production 51,920 TJ.

Switzerland 41,277 km², 8.06M

2000: The installed capacity for GHPs is 500 MW_{th}, and the annual production 1,980 TJ.

2005: GHPs account for 532.4 MW_{th} and 2,854 TJ/yr.

2010: 1,017.1 MW_{th} and 6,602 TJ/yr for GHP.

2015: 1,684 MW_{th} and 10,897 TJ/yr for GHP.

Turkey 783,562 km², 76.67M

2000 No data; 2005 No data; 2010: For GHPs, 38 MW_{th} and 536.5 TJ/yr is reported.

2015: 42.8 MW_{th} and 960 TJ/yr for GHPs.

Ukraine 603,550 km², 44.29M

2000 No data; 2005 No data; 2010 No data; 2015 No data.

United Kingdom 243,610 km², 63.74M

2000: The installed capacity of GHPs is 0.63 MW_{th} and the annual production 2.73 TJ.

2005: Estimated are 10.2 MW_{th} installed capacity and an annual production of 45.6 TJ.

2010: 181.50 MW_{th} and 753.40 TJ/yr for GHPs.

2015: 280 MW_{th} and 1,800 TJ/yr for ground source heat pumps.

Chapter 8

Policy and Regulation
for Geothermal Energy in the EU

Luca Angelino

European Geothermal Energy Council, Brussels
l.angelino@egec.org

This chapter presents the policy and regulatory framework relevant for geothermal energy. It provides an overview of the key legal challenges, of the European Union (EU) energy and climate framework, and of the evolving national financial incentives available. The analysis introduces to the licensing system in select countries and presents the main regulatory instruments used to ensure that geothermal projects are compatible with the environment. The EU energy and climate objectives can, in many ways, support the sector: in this context, the RES Directive has been the most important piece of legislation addressing many nontechnical barriers for geothermal energy. But policy and regulatory instruments evolve rapidly: with the gradual phase-out of feed-in tariffs and with more constraints on public budget, the future of the sector may largely depend on innovative financial mechanisms coupled with more market-based incentives.

8.1. Introduction

The state intervenes in the private domain with the primary objective to promote the general interest. This is justified by the very fact that "we live in a world of finite resources, in which the pursuit of self-interest often fails the individual and causes harm to others" (Orbach, 2012, p. 4). With a varying degree of detail, public authorities tend to regulate every economic activity, which inevitably affects the behavior of many industries, including those involved in geothermal energy. To give an example, regulations put in place to preserve the quality of the environment may limit the location, require specific procedures, and ultimately have an effect on time and cost of a geothermal project.

The objective of this chapter is to introduce the complex and evolving policy and regulatory framework[1] relevant to geothermal energy in Europe. The analysis covers all of the various geothermal technologies described in other chapters of this book and has a focus on the European Union (EU) legislation and its implementation. Indeed, nowadays, it remains difficult to fully understand the legal system for geothermal energy in a given country without some acquaintance within the overarching EU framework. To this end, it may be useful to clarify some preliminary principles governing the relation between the EU and its member states. First, the competences between the two levels are defined in the Treaty on the Functioning of the European Union. In areas like energy and the environment, where the competence is shared, the EU can legislate when its action is considered to be more effective than the action taken at national, regional, or local level (principle of proportionality). Second, the legal basis of the EU action determines the margin for maneuver of national governments. When the objective is the removal of obstacles to trade, the EU adopts regulations that are simultaneously, automatically, and uniformly binding in all national legal systems. When the objective is different (e.g., environmental protection), the EU tends to adopt directives establishing a number of objectives that member states have to achieve. They have, however, the flexibility to devise the most appropriate means in line with national preferences. In this context, it is worth highlighting that under no circumstances national rules can contradict EU legislation;

---

[1]Throughout this chapter, I will use several times the terms "policy", "legislation", and "regulation". I refer to "[public] policy" as the action (or nonaction) of governments in order to achieve certain goals, to "legislation" as the set of primary laws or relevant legislative provisions concerning a given domain, and to "regulation" as the rules, administrative in nature, based on and enacted to implement a specific piece of primary legislation.

if that happens, EU law will prevail (principle of supremacy). Third, EU legislation goes beyond the EU borders. In the framework of the functioning of the European Economic Area, EU rules can apply to Iceland, Norway, and Lichtenstein. Additionally, they can also apply to other countries (e.g., Switzerland) through bilateral agreements.

Once the relation between the EU and the national level of governance is clarified, it is important to remind that national competences can be further devolved to regional and local authorities depending on the degree of self-governance in each country. And it is precisely because of the interdependence between many levels of governance that the present analysis cannot be exhaustive. Yet, it will provide background information for further research.

The chapter is organized as follows: Section 8.2 provides an overview of the key legal issues for the sector and a brief presentation of the licensing systems in Italy, Germany, France, and Hungary; Section 8.3 takes a closer look at the EU climate and energy framework; Section 8.4 presents an overview of the most common mechanisms put in place by governments for supporting geothermal energy and an assessment of the expected evolution in this field; finally, Section 8.5 summarizes the main conclusions.

8.2. Key Regulatory Issues

A geothermal system is developed in several phases. As illustrated in Figure 8.1, a simplified way to classify the different steps in a deep geothermal project is as follows: (a) exploration; (b) resource development; (c) construction; and (d) commissioning and operation.

Each of these phases requires one or more permits and the compliance with a range of national and local rules. The whole set should be as transparent and balanced as possible in order to ensure, simultaneously, the sustainable use of the resource, confidence in the technology, and investment security. Several studies (Goodman *et al.*, 2010; GEOELEC, 2013; GEODH, 2014; REGEOCITIES, 2015) have assessed the most relevant regulatory issues affecting the geothermal sector, which can be classified as follows: (a) definition, classification, and resource ownership; (b) licensing and

	Year 1	Year 2	Year 3	Year 4	Year 5	Year 6	Year 7	Year 8
Exploration	Exploration & test drilling							
Resource development			Drilling					
Construction						Engineering & construction		
Commissioning & operation								O&M

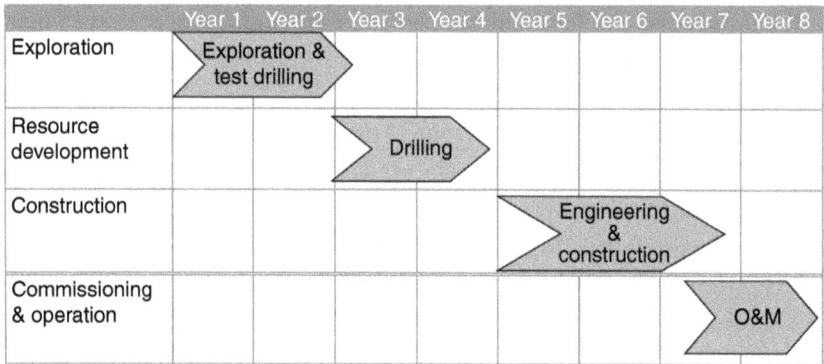

Figure 8.1   Phases of a deep geothermal project.
Source: GEOELEC (2013, p. 41).

authorizations; (c) sustainability; (d) spatial planning; and (e) access to the grid. The following sections present each of these categories.

8.2.1. Definition, Classification, and Resource Ownership

The definition, classification, and ownership of geothermal resources affect many key aspects of regulation in this field. In Europe, Directive 2009/28/EC (RES Directive) provides a legally binding definition according to which "geothermal energy" means energy stored in the form of heat beneath the surface of solid earth (Art. 2). Geothermal is therefore to be considered as a renewable energy source of its own kind and any national or local regulation should be in line with the above overarching definition. In practice, however, geothermal resources are still defined in a several ways: as mineral, water, or groundwater, heat and, more rarely, as a *sui generis* resource.

Besides that, a classification between different types of resources is useful for determining the regulatory approach to the various categories of geothermal systems. The EU-funded project GeoThermal Regulation-Heat (GTR-H) has proposed that a single depth limit ought to be used to define and differentiate geothermal resources depending on country-specific geological conditions. For example, deep geothermal resources could be defined as occurring below depths of 200–500 m and shallow resources as those located above the chosen threshold (Goodman *et al.*, 2010). In this regard, the RES Directive refers to "shallow geothermal" in relation to training and certification, but does not propose any further distinction. Without an EU-wide approach, member states use various methods (see Box 8.1 for Italy, France, Germany, and Hungary).

Another basic but essential legal issue is the resource ownership definition. Three situations can be found. The first is when the geothermal resource belongs to the state and a number of permits are needed for its use. This is the case in most of the European countries with plants in operation and the most desirable of the options. A second case, more typical of common law systems, is when the resource belongs to the owner of the surface estate; this situation is particularly unfavorable for deep geothermal projects, where multiple owners are concerned. The third and worst case is found is some juvenile markets where there are no specifications about ownership. Traditionally, a first come–first served approach is in place, unless priority is given by law to a specific use (GEOELEC, 2013).

8.2.2. Licensing and Authorizations

Public intervention is often perceived as a way to control and restrict behavior, but in several cases, it is an essential enabler. The mere existence of a licensing or authorization procedure is an example of how governments can enable an economic activity. In the geothermal sector, a true license provides exclusive rights within a certain area and for a given time period, thereby ensuring investment security. Additionally, a licensing regime tends to clarify issues such as who is eligible to obtain a permit, who are the licensing authorities, how many steps and the time the process involves, the exact time period for which a license can be obtained and extended, if royalties are required, and under what parameters.

The type of permits a project developer must obtain and the respective procedures to follow depend primarily on the definition and classification of geothermal resources (see Section 8.2.1). Being an underground resource, the administrative procedures relevant for geothermal stem from a long history in mining and are in many cases part of a wider legal framework intended for coal, hydrocarbons, etc. In the vast majority of European countries, the licensing regime for deep geothermal consists of a two-step process requiring an exploration and a production license. In addition, a number of other permits could be required during these two phases. As we shall see in Section 8.2.3, these permits allow the public authorities to ensure that the project is performed in a safely and environmentally sound way and fulfills all public participation and consultation requirements. Box 8.1 provides a brief overview of the licensing and authorization regimes in Italy, Germany, France, and Hungary.

Box 8.1 Overview of licensing and authorization regimes in selected countries.

Italy

A dedicated framework for geothermal energy was established in 1986 and has been reshaped in 2010. Geothermal energy is defined as a mineral resource distinguished between high-enthalpy resources (over 150°C), medium-enthalpy resources (between 150°C and 90°C), and low-enthalpy resources (below 90°C). High-enthalpy resources are of national interest together with areas able to provide plants with a capacity of at least 20 MW. Other types of resources belong to the regions (Legislative Decree No. 22/2010). The competence to assess and issue exploration and development licenses is devolved to the regions, but for pilot projects of maximum 5 MW aiming at the total reinjection of the geothermal fluid into the same aquifer. Legislative Decree. No. 28/2011 implementing the RES Directive defines these types of resources to be of national interest and provides that the competent authority is the Ministry of Economic Development, in agreement with the Ministry of Environment and with the region concerned. Pilot projects are also exempt from paying the royalties applying to geothermal power plants €0.0013 per kW/h to the interested municipalities and €0.00195 per kW/h to the region.

(Continued)

Box 8.1 (*Continued*)

In order to make use of the geothermal resources of national interest, an exploration and a production license are required. The exploration license gives the holder the exclusive right for exploration within a defined area and a specified time period. The area cannot exceed 300 km^2 and the term can be set up to 4 years and can be extended up to 2 years. The development license is granted up to 20 years and extendible for 10 more years.

Additionally, the Italian regulation further defines "small local uses" of geothermal resources belonging to the regions as follows:

- All closed-loop systems.
- Open loop systems with a thermal capacity of maximum 2 MW and with wells located up to 400 m depth.

A simplified procedure for authorizing closed-loop systems was announced (Art. 7 Legislative Decree 28/2011) but, as of 2016, is not yet in place. In the absence of national regulation, there are only few regions that have adopted their own regulations; in the others the regulatory gap is still evident (Abate *et al.*, 2014, p. 100).

Germany

In Germany, geothermal energy is defined as a mineral resource. While for shallow geothermal projects up to 100 m depth simplified procedures are in place in some states, exploration and development of deep geothermal is subject to licenses issued by the state mining authority. A prerequisite is to gain the consent of eventual landowners (Fraser, 2013, p. 31).

As far as groundwater protection is concerned, water legislation at state level applies. As a result, there are different water regulations in force. In addition, the developer must submit a number of short-term and long-term operational plans. For geothermal projects involving drillings over 1,000 m depth, located in protected areas, or involving water abstraction, the main operational plan must obligatory include an environmental impact assessment (Fraser, 2013, p. 33, 35; Bradbrook and Rønne, 2014).

(*Continued*)

Box 8.1 (*Continued*)

The exploration license gives the geothermal developer the exclusive right for exploration within a defined area and for the specified resources. It can be granted up to 5 years and may be extended for 3 years. The development license includes the authorization for work and can be granted for a period not exceeding 50 years.

France
Geothermal resources fall under the mining law, which provides for a two-step process requiring an exploration and a production license. In addition, and in order to prevent pollution and industrial risks, a specific work permit is necessary for each operation that is performed in relation to the underground. The exploration license gives the licensee exclusive rights for exploration within a defined area and for a period up to 5 years, extendible twice. During this period, the holder of the exploration license is the only one eligible for a production license for the same area. The subsequent production license gives an exclusive right to use the geothermal resource for energy purpose up to 50 years. This license is extendible for 25 years. The applications are submitted to the Ministry of Mines and examined by a representative of the state (the *Préfet*) at the local level. This authority is also in charge of organizing a public inquiry (Fraser, 2013, pp. 4–7).

It is worth noting that the French regulation provides simplified procedures for specified shallow geothermal projects. Decree No. 2015-15 of January 8, 2015 has established an online notification procedure for projects qualified as being of "minimum significance", to be understood as being unlikely to have adverse environmental impacts. The simplified procedure applies to closed-loop and open-loop systems of less than 200 m depth and whose thermal capacity is lower than 500 kW. An additional requirement for open-loop systems is that the flow rate is lower than 80 m^3/h and that the groundwater is reinjected into the same aquifer.

(*Continued*)

Box 8.1 (*Continued*)

These projects must be located in preidentified areas (so-called green and orange zones, with the latter requiring a certificate from an authorized expert). The simplified procedure does not apply if the activities are located in so-called red zones and for systems not falling under the above categories. In this case, an authorization reviewed by the relevant authorities must be obtained.

Hungary

In Hungary, different rules apply depending on the following two parameters:

- Whether the development requires water abstraction; and
- The depth of the targeted geothermal resource.

First, when water abstraction is involved between a depth of 20 and 2500 m, provisions of the environmental and water management legislation are to be considered. In this case, the relevant regional governmental office is the competent authority for issuing a license for thermal groundwater, extraction and drilling.

Second, exploration and development of geothermal energy below 2,500 m takes place in the frame of a concession (Art. 49 of the Mining Act XVVIII of 1993). In this case, the responsible minister (minister of economic development) announces a public tender published in the *Official Journal of the European Union*. The tender is evaluated within 90 days, after which a contract is signed with the winner and a concession fee has to be paid. The concession is granted for a period up to 35 years and may be extended once by not more than half of the term. The exploration cannot be longer than 4 years within the concession period and can be extended twice.

Different permits are needed. First, in the preparation phase, the most important is the environmental permit, which is subject to an environmental impact assessment and compulsory in case groundwater abstraction exceeds 5 million m^3/year, or reinjection of 3 million m^3/year and in all cases where groundwater

(*Continued*)

Box 8.1 (*Continued*)

extraction from karstic aquifers exceeds 500 m³/day or 2000 m³/day from porous aquifers. This license is a prerequisite to start all construction/drilling operations. Second, in the development phase, wells up to 2,500 m require a water construction license, while other permits are required from the Energy and Public Utility Regulatory Authority for the construction of the plant as well as of a distribution network. Third, during the operation phase, the water operation license and an operation license for the plant are necessary (Angelino *et al.*, 2016).

In Hungary, there is no simplified procedure for small-scale geothermal heat pump projects. Closed-loop systems up to 20 m depth require neither a license nor a notification to the authority. A register of these types of system does not exist in the country.

While the licensing regime is a key enabler for the geothermal business, it should, however, be very well regulated. As a matter of fact, in some countries, the right to use the geothermal resource is not clearly exclusive, while in other cases complex, long, and sometimes unnecessary procedures represent a significant nontechnical barrier for geothermal developers. Delays, for example, can provoke uncertainty and lead to higher risks due to which investors require higher returns. For a capital-intensive technology, a one stop-shop process is desirable for each phase of the project (GEOELEC, 2013, p. 77).

In this regard, the RES Directive aims to improve the legal framework for RES projects. To break with the past, Article 13 requires member states to streamline and rationalize the administrative procedures and to clearly define and coordinate the respective responsibilities of national, regional, and local administrative bodies. Besides, timetables for determining planning and building applications should be transparent: comprehensive information and assistance to applicants should be made available at the appropriate administrative level. Furthermore, it specifically requires taking into account the particularities of individual technologies and encourages the practice of "simple notification" for smaller projects. Good practice in this sense for shallow geothermal systems is found in France and is presented in Box 8.1.

8.2.3. Sustainability

Negative environmental impacts associated with geothermal energy are minor, especially if compared with conventional fossil fuels and nuclear power plants. As a matter of fact, a geothermal plant is located right above the resource and does not imply mining, processing, transporting the fuel over great distances, and combustion. Furthermore, the visual and land use impact can be negligible. However, as for every industrial activity, some potential and adverse effects exist such as some forms of gaseous emissions, induced seismicity, ground subsidence, noise during the construction phase, and temperature anomalies in the subsurface and the groundwater. These potential impacts vary depending on the geological settings as well as on the size and type of application. Anyhow, they can be avoided thanks to sound practice, technology development, and compliance with environmental regulations.[2]

A high level of protection and improvement of the quality of the environment is one of the objectives of the EU (Art. 3 Treaty on European Union); in consequence, the body of secondary legislation is very comprehensive. For deep geothermal, the most relevant pieces of legislation are the following:

- Directive 2011/92/EU on the assessment of the effects of certain public and private projects on the environment (Environmental Impact Assessment [EIA] Directive)[3];

---

[2]For an analysis of sustainability for deep geothermal see Tester *et al.* (2006, 8-3–8-19)], Goldstein *et al.* (2011, pp. 418–420), and GEOELEC (2013, pp. 72–77), while for shallow geothermal energy see the contribution of Rybach and Sanner for this book and Hänlein *et al.* (2013). The above studies give access to a wider literature on the topic.

[3]As amended by Directive 2014/52/EU.

- Directive 2001/42/EC of the European Parliament and of the Council of June 27, 2001 on the assessment of the effects of certain plans and programs on the environment (SEA Directive); and
- Directive 2000/60/EC establishing a framework for community action in the field of water policy.

The EIA of certain projects before their authorization is universally acknowledged as a basic instrument for the protection of the environment and the application of appropriate preventive measures (Thieffry, 2015, p. 639). The EIA and the SEA directives align the EU *acquis* with a number of international treaties, chiefly the 1991 UNECE Convention on Environmental Impact Assessment in a Transboundary Context (Espoo Convention) and the 1998 UNICE Convention on Access to Information, Public Participation in Decision-Making and Access to Justice in Environmental Matters (Aarhus Convention). Despite a common EU framework, the rules applying to the deep geothermal sector may differ from one country to another. As a matter of fact, deep geothermal drilling is among the activities listed under Annex II of the directive and for which are the national authorities to determine when an EIA in the sense of the directive is required (Art. 4(2)). Table 8.1 presents the steps and the options to follow for such determination.

If an EIA is eventually required, a number of common minimum procedures set out in articles 5–10 and Annex IV EIA Directive are to be followed. The process consists of five main phases:

1. The project developer presents a detailed report and a nontechnical summary containing a description of the physical characteristics of the whole project, including the impact on the environment, the mitigation measures envisaged, etc.
2. A minimum 30-day public consultation takes place, which is a key part of the process to ensure the right of participation.
3. The authorities examine the report and the relevant information received in the framework of the consultation.

Table 8.1   Determination of when a deep geothermal project requires an EIA in the sense of Directive 2011/92/EU.

1: During the application for exploration and/or production license, the developer is asked to provide information on the characteristics of the project and its likely significant effects on the environment, including measures envisaged to avoid or prevent eventual adverse effects.

2: National rules can determine when an EIA is required on the basis of thresholds/criteria or on a case-by-case examination. Countries such as Hungary and Germany have set strict criteria to establish which geothermal drilling projects must go through an EIA in the sense of Directive 2011/92/EU (see Box 8.1).

3: In a case-by-case examination based on the information submitted by the developer (see point 1), member states have to take into account some minimum criteria set out in Annex III EIA Directive. The decision on the screening must be published and taken as soon as possible within 90 days from the date on which the developer has submitted all the information.

4: In compliance with Directive 92/43/EEC (Habitats Directive), an EIA is always required if a project is developed within a "natural habitat types of Community interest".

---

4. The authorities issue a reasoned conclusion on the direct and indirect impacts on the following factors:

— Population and human health;
— Biodiversity;
— Land, soil, water, air and climate;
— Material assets, cultural heritage;
— The landscape.

5. Such reasoned conclusion is integrated in the development consent decision of the competent authority.

Only following those steps, and provided there are no complications, the developer can be entitled to proceed with the project. All in all, it is observed that going through such a process may be a very demanding exercise, especially for small and medium-sized enterprises.

The Strategic Environmental Assessment (SEA) Directive extends the evaluation of the environmental impact to certain plans and programs, for example, an urban or energy plan including a geothermal DH project, prepared or adopted by a national, regional, or local authority and required by legislative, regulatory, or administrative provisions. The EIA and SEA procedures are very similar and both foresee the consultation of the public and of other member states if a project, a plan, or a program is likely to have significant effects on the environment beyond the jurisdiction of the territory in which it is being prepared.

As far as the protection of water quality is concerned, Directive 2000/60/EC (Water Framework Directive) represents the main piece of legislation on the matter in Europe. It requires member states to implement the necessary measures to prevent the deterioration of the status of all bodies of surface and groundwater. Concerning the latter, for which pollution prevention and quality monitoring and restoration are more difficult mostly due to its inaccessibility, the directive takes the precautionary approach and establishes a general prohibition on direct discharges to groundwater. An important exemption is, however, provided for geothermal energy: in this case, member states are given the option to authorize reinjection into the same aquifer, provided it does not compromise the environmental objectives of the directive (Art. 11). It is therefore within the competence of the national governments to decide as to whether reinjection of the geothermal fluids is allowed or even required.

8.2.4. Spatial Planning

The planning of local infrastructure plays an important role for geothermal heating and cooling systems. For example, the very technical feasibility of geothermal heat pump systems may depend on the interaction with underground infrastructure such as parking areas and communication and transport systems. For this reason, it is essential to know the position and the dimensions of this infrastructure to avoid undesirable interference and ensure that the systems can be installed in the planned position for a long period of time.

Here, local rules in terms of underground planning play a very important role (REGEOCITIES, 2015).

Regarding new geothermal district heating (DH) systems, it is the rigidity of local plans for heating and cooling, which once implemented are difficult to alter, which may represent a significant barrier. For this reason, the RES Directive recommends member states to encourage local and regional administrative bodies to include heating and cooling from RES in the planning of city infrastructure (Art. 13). This provision has resulted in some positive changes, for instance in Italy where Legislative Decree 28/2011 imposes on municipalities with more than 50,000 inhabitants the requirement to draft DH development plans. In this regard, it should be mentioned a requirement from Directive 2012/27/EU (Energy Efficiency Directive or EED) to carry out comprehensive assessments and a cost–benefit analysis of the potential for the application of high-efficiency cogeneration and efficient DH and cooling as a basis for sound planning (Art. 14). When a potential for the construction of the related infrastructure is identified and its benefits exceed the costs, adequate measures to accommodate its realization should be put in place. It is still to be seen, however, whether the actual implementation of the above provisions can concretely have an impact on the development of geothermal energy as a renewable and efficient source for DH.

8.2.5. Grid Access

For geothermal power, connection and access to the grid is the last 28 step before remuneration. Given the former monopolistic structure 29 of the electricity market, grid connection and access for new and 30 especially renewable power plants has not always been obvious. This 31 is why there was a need for clear and nondiscriminatory rules: the 32 RES Directive has addressed this issue by requiring priority or guaranteed 33 access to the grid for renewable electricity (Art. 16). The above provision will be in force until 2020 and constitutes specific legislation for the connection and dispatching of electricity generating installations. The issue, however, is mainly covered by the Electricity Directive and Regulation, which set the general rules for the electricity sector in the EU (see Section 8.3.2).

8.3. The European Union's Energy and Climate Framework

The above review of the key regulatory issues for geothermal energy has already shown the strong impact of EU legislation on the sector. It is, therefore, worth having a closer look at the EU framework, this time limiting the analysis to the energy and climate field, to see how it is contributing to the market development of the different geothermal technologies.

8.3.1. Objectives and Dimensions

Energy has not always been an official competence of the EU institutions. It was the response to critical supranational issues such as climate change and security of supply that made the development of a more comprehensive EU energy policy indispensable. The 2009 Lisbon Treaty formalized this development with the addition of a specific title on energy (Title XXI) in the Treaty on the Functioning of the European Union. Before this formal recognition, however, "[European] policy-makers borrowed legal competence from the economic and environmental parts of the treaties to justify proposing and passing energy measures" (Buchan, 2009, p. 7).

The EU objectives in this policy area are listed in Art. 194 and are the following:

1. To ensure the functioning of the energy market.
2. To ensure security of energy supply.
3. To promote energy efficiency and energy saving and the development of new and renewable forms of energy.
4. To promote the interconnection of energy networks.

The same article states that the choice of the energy mix lies with the member states but makes clear their obligation to integrate environmental considerations into energy policy. Based on these new provisions, in 2015 the European Commission, the EU's executive body, has reorganized all the EU actions in the field in a framework strategy toward the establishment of a "resilient Energy Union with a forward looking climate policy" (European Commission, 2015;

Delbeke *et al.*, 2015, p. 64). The strategy is being built around the following five dimensions:

(a) Energy security, solidarity, and trust.
(b) A fully integrated European energy market.
(c) Energy efficiency contributing to moderation of demand.
(d) Decarbonizing the economy.
(e) Research, innovation, and competitiveness.

The next sections discuss the matters relevant to geothermal energy in all of the five dimensions of the EU Energy Union.

8.3.2. Security of Supply and Internal Energy Market

Regarding the first dimension of the energy union, concerns over security of supply stem from the traditional reliance of most EU member states on imported fuels. Just to give an example of the impact of such dependency, the EU external energy bill amounted in 2013 to some €400 billion, representing more than €1 billion per day (European Commission, 2014). EU legislation in this field integrates the wider OECD agreements developed following the oil crises in the 1970s and consisting of emergency response measures and prevention strategies. Only the more recent Russia–Ukraine crises and the serious vulnerability of Central and Eastern member states resulted in a stronger link between security of supply, on the one hand, and energy efficiency and fuel switch to renewable energy, on the other hand.

As far as the second dimension is concerned, building on the single market program, the European Commission has been working since 1989 on the not yet completed liberalization of the electricity (and gas) markets. Through three different packages of measures (in 1996, 2003, and 2009), the traditional monopolistic market structure in place for decades and characterized by vertically integrated suppliers has been formally removed. One of the most important provisions of the 2009 Electricity Directive is indeed the so-called ownership unbundling, i.e., the separation of power generation from supply networks in order to give every user a fair, transparent, and equal access to a natural monopoly, as it is the electricity grid. In other words, the former monopolists were forced to sell their networks or to put them under a completely independent management. In November 2016, the European Commission proposed a new reform aiming to adapt the electricity market design to the growing share of variable electricity generation from wind and photovoltaic. Amongst other things, the proposal opens up balancing markets to all market participants, including independent aggregators and demand response.

The main consequence of the liberalization of the electricity market for geothermal was the possibility for new players to enter into the business. In Italy, the liberalization of the electricity market was complemented by the liberalization of the geothermal resources through Legislative Decree 22/2010, which has resulted in 50 new exploration licenses granted, many of which outside the traditional geothermal region of Tuscany.

8.3.3. Decarbonization Policy, the Renewable Energy Directive, and Energy Efficiency Legislation

In order to contribute to the global efforts to mitigate climate change, the EU has the objective of reducing greenhouse gas emissions by 80–95% by 2050 compared to 1990. According to the Intergovernmental Panel on Climate Change, this is in line with the necessary reductions by developed countries as a group in order to achieve an overall global reduction of 50% by mid-century.[4]

Compared with security of supply and energy market, EU efforts to reduce greenhouse gas emissions have a way more direct impact on the geothermal sector. As a matter of fact, the so-called EU 20-20-20 targets endorsed by EU heads of state and government have triggered a renewed interest in geothermal energy technology. The 20-20-20 goals, which are headline targets of the European 2020 strategy for growth, are the following:

- Reduction of at least 20% in greenhouse gas (GHG) emissions compared to 1990 levels.

---

[4]It is worth noting that such assessment is based on the prior objective to limit the global average temperature increase to 2°C above preindustrial levels, while the 2015 Paris agreement includes a new objective to limit the increase to 1.5°C.

- 20% of the final energy consumption to come from renewable sources.
- Improvement of energy efficiency by 20% compared to 2007 projections.

Additionally, a new framework was adopted by EU leaders in October 2014, with the following three key targets for the year 2030:

- At least 40% cuts in greenhouse gas emissions (from 1990 levels).
- At least 27% share for renewable energy binding on the EU level only.
- At least 27% improvement in energy efficiency binding on the EU level only.

A set of instruments contribute to the delivery of the expected results for 2020 and 2030. The EU-wide emission trading system (ETS) and national targets for GHG emissions reduction in nonETS sectors[5] are very important but are not analyzed here because of their limited direct impact on geothermal technologies. The following sectorial directives are instead the most important for the geothermal sector:

1. Directive on the promotion of the use of energy from renewable sources (2009/28/EC), setting national binding targets until 2020.
2. Directive on energy performance of buildings (2010/31/EU), setting minimum requirements for new and refurbished buildings.
3. Directive on energy efficiency (2012/27/EU) promoting renovation and energy savings through obligations and behavioral changes.
4. Directives on eco-design requirements (2009/125/EC) and energy labeling (2010/30/EU), promoting efficiency of products.

"In 2016, the European Commission proposed a revision for the post-2020 period of the directives on renewable energy, energy performance of buildings, and energy efficiency". The following

---

[5]The EU Emissions Trading Systems (ETS) covers electricity and industrial installations above 20 MW, and those not covered by the ETS are buildings, services, and small industries, and land use, land-use change and forestry (LULUCF).

section present the directives in place at the time of writing (December 2016).

8.3.3.1. The 2009 renewable energy directive

The RES Directive was designed to ensure the achievement of the 2020 renewable energy targets. As shown earlier, it addresses a number of key barriers for the deployment of geothermal such as lack of a widely accepted definition, administrative barriers, spatial planning, and grid access (see Sections 8.2.1, 8.2.2, 8.2.4, and 8.2.5). Moreover, it translates the EU target into legally binding national targets (Art. 3 and Annex I). In order to provide more flexibility to member states, cooperation mechanisms such as joint projects, joint support schemes, and statistical transfer are included (Articles 6–10). In addition, the directive requires governments to submit national renewable energy action plans (NREAPs) including a qualitative analysis relating to the planned policy measures and projections for each technology in electricity, heating and cooling, and transport (Art. 4).[6] Member states report on their progress every two years.

Tables 8.2–8.4 show the trend for geothermal power, heat, and geothermal heat pump technology, including the projections until 2020 as provided by member states in their NREAPs.

Noteworthy for the sector is a new methodology, briefly summarized in Box 8.2, established for accounting the contribution of electric heat pumps, including geothermal heat pumps, toward the renewable energy targets. This was seen since the beginning as a significant implementation challenge given the diversity of heat pump applications and the scarcity of statistical data (Hodson *et al.*, 2010), which had sometimes acted as a barrier for the development of the technology. As it is possible to note in Table 8.4, not all member states report the breakdown by the different types of heat pumps; therefore, this addition is not always contributing to quantify the energy and thereby the impact of the shallow geothermal sector.

---

[6]Technology-specific trajectories and sectorial and intermediate targets are not binding. Only if a member state does not meet the overall target in 2020 it will be subject to financial fees in line with EU law.

Table 8.2   Trends in the power sector in the EU (MWe).

Country	2010	2015	2020
Italy	754	915.5	920
Germany	10	36.6	298
Greece	0	0	120
France	26	17.1	80
Portugal	25	23	75
Hungary	0	0	57
Spain	0	0	50
Ireland	0	0	5
Czech Republic	0	0	4.4
Croatia	N/A	N/A	10
Slovakia	0	0	4
Belgium	0	0	3,5
Romania	0	0.05	
Austria	1	1	1
EU	816	993.2	1627.9

Source: National Renewable Energy Action Plans, EGEC Market Report 2015.

Table 8.3   Trends in geothermal heat production in the EU (ktoe).

Country	2010	2014	2020
Germany	57.1	91	686
France	98.2	125.7	500
Hungary	98.4	124.5	357
Italy	139.3	129.6	300
Netherlands	7.6	35.9	259
Poland	13.4	20.2	178
Slovakia	5	4.2	90
Greece	16	11.7	51
Austria	20.5	19.4	40
Portugal	1	1.3	25
Slovenia	26.3	30.9	20
Czech Rep	0	0	15
Croatia	6.8	10.7	15.7
Spain	16	18.8	9.5
Bulgaria	32.7	33.4	9
Belgium	2.1	1.4	5.7
Lithuania	2.3	0.9	5
Romania	22.1	25.1	80
UK	0.8	0.8	0
Denmark	2.5	2	N.A.
Cyprus	0.8	1.6	N.A.
EU	568.9	689.1	2630.2

Source: EUROSTAT Shares 2014, National Renewable Energy Action Plans.

Table 8.4   Trends in production from geothermal heat pumps in select EU member states.

Country	2010	2014	2020
UK	21.7	56.6	953
Sweden	N.A.	803.3	815
France	217.1	261.6	570
Italy	44.2	70.8	522
Germany	246.2	334	521
Netherlands	52.1	81	242
Denmark	56.2	71.8	199
Hungary	N.A.	N.A.	107
Greece	N.A.	N.A.	50
Spain	N.A.	16.4	40,5
Slovenia	N.A.	N.A.	38
Austria	N.A.	N.A.	26
Slovakia	N.A.	N.A.	4
Romania	N.A.	N.A.	8
Finland	N.A.	133.8	N.A.
Czech Republic	26.5	41.8	N.A.
Estonia	N.A.	21.1	N.A.
Poland	3.1	8.4	N.A.
Hungary	N.A.	5	N.A.
Luxembourg	N.A.	0.5	N.A.

Source: EUROSTAT SHARES 2014, National Renewable Energy Action Plans. Countries not reported in the table have not reported the breakdown of heat pumps by source.

Box 8.2 Methodology to calculate renewable energy from heat pumps in the RES Directive.

The energy used to drive heat pumps should be deducted from the total usable heat. Only heat pumps with an output that significantly exceeds the primary energy needed to drive it should be taken into account. Accordingly, the quantity of heat to be considered as energy from renewable sources for the purposes of this Directive shall be calculated in accordance with the methodology laid down in Annex VII:

$$\text{ERES} = Q_{\text{usable}} * (1 - 1/\text{SPF})$$

(*Continued*)

Box 8.2 (*Continued*)

where

ERES is the amount of energy captured by heat pumps to be considered energy from renewable energy sources for the purposes of this Directive;
Q_{usable} *is the estimated total usable heat delivered by heat pumps fulfilling the primary energy efficiency criterion, and*
SPF is the estimated average seasonal performance factor for those heat pumps.

In line with Annex VII, in March 2013 the European Commission adopted a decision (C(2013) 1082 final) establishing guidelines on how member states could estimate the two parameters Q_{usable} and SPF.
If no better data from actual measurements are available, Q_{usable} should be calculated as follows:

Q_{usable} *estimated total usable heat*
H_{HP} *full-load hours of operation*
P_{rated}: *capacity of heat pumps installed*

The guidelines provide default values for H_{HP} and SPF for three different climate zones (cold, average, and warm). The guidelines may be revised and complemented by the Commission if statistical, technical, or scientific progress necessitates it.

8.3.3.2. Geothermal, buildings and energy efficiency legislation

Buildings are responsible for nearly 40% of final energy consumption in the EU (European Commission, 2011) and have a large potential for energy savings and carbon emissions reduction, including through the application of geothermal technologies. At the EU level, a series of measures have been adopted to improve the energy performance of buildings and products and to integrate renewable energy into new buildings and existing buildings subject to major renovation.

Table 8.5   Main measures and timetable related to geothermal in buildings.

Date	Provision
Since 2014	Member states to renovate each year an average 3% of the public building stock owned by central governments (Article 5 Energy Efficiency Directive).
Since 2015	Member states to introduce, where appropriate, measures to set minimum levels of RES, which should be used in buildings or equivalent supporting measures (RES Directive) and intermediate targets for improving the energy performance of new buildings (Art. 9.3 (b) EPBD).
September 26, 2015	Energy label for brine-to water heat pumps A++ to G introduced (substituted by a new label ranging from A+++ to D in 2019).
December 31, 2018	All new buildings owned or occupied by public authorities shall be nearly zero-energy buildings (Art. 9.1 EPBD).
December 31, 2020	All new buildings (including private buildings) shall be nearly zero-energy buildings (Art. 9.1 EPBD).

The most relevant measures having a potential positive impact on geothermal heating and cooling technologies are reported in Table 8.5 and discussed below.

Directive 2010/31/EU (Energy Performance of Buildings Directive or EPBD) requires member states to set primary energy requirements for new and existing buildings undergoing major renovation. These requirements are to be reviewed every five years and should be calculated through a cost-optimal methodology taking into account certain elements, including the thermal characteristics of a building. For new buildings, high-efficiency alternative systems, including geothermal heat pumps and district or block heating or cooling, need to be assessed before construction starts. In addition, the EPBD and introduces in EU law the ambiguous concept of "nearly zero-energy building (NZEB)", which is "a building that has a very high energy performance, whose very low amount of energy required should be covered to a very significant extent from energy from renewable sources produced on-site or nearby" (Art. 2). These provisions are linked with the RES Directive, according to which member states should, in their building regulations and codes, or by other means of equivalent effect, require the use of minimum levels of RES in new buildings and existing buildings undertaking major renovation. As a result of the above, member states have laid down their own detailed NZEB definition. Regarding renewable energy, most member states require a renewable energy share of the primary energy or a minimum renewable energy contribution in $kWh/m^2 \cdot year$, while others use indirect requirements, such as a low nonrenewable primary energy use that should only be met with the use of renewables (Erhorn and Erhorn-Kluttig, 2016).

Relevant to geothermal heat pump technology only are both the eco-design and energy-labeling regulations. Eco-design aims to improve the energy and environmental performance of products throughout their life cycle, while energy-labeling requirements aim to provide the consumer with information about the environmental performance of products. New eco-design requirements and energy labels for space heaters and combination heaters entered into force in September 2015. An energy label for brine to water heat pumps is established in two phases: the first entered into force in 2015 ranges from A++ to G, while a second ranging from A+ + + to D will be introduced in 2019. As geothermal heat pump is among the few technologies to achieve the highest class, this instrument has a significant potential to increase awareness.

8.3.4. Research and Innovation

The decarbonization and energy efficiency dimensions analyzed in the foregoing section have an enormous impact also on the research and innovation (R&I) dimension of the Energy Union. Together with national programs, since the 1990s, the EU has contributed to R&I in geothermal by financing resource assessment and a number of demonstration projects, the most famous of which is the flagship enhanced geothermal systems (EGS) project in Soultz-Sous-Forêts situated along the French–German border. Figure 8.2 illustrates the historic allocation of EU funding for R&I to geothermal until 2012, while a number of new projects have been initiated under the current

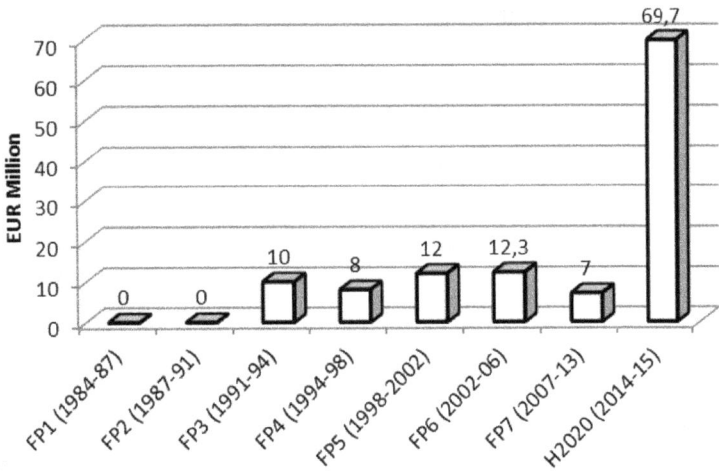

Figure 8.2 Historic EU R&I support for geothermal energy.
Source: European Commission.

Framework program "Horizon 2020" running from 2014 through 2020.

In terms of governance, R&I efforts to accelerate the deployment of cost-effective low-carbon technologies have been organized under the so-called strategic energy technology (SET) plan. Established in 2008, it is based on a three pillar implementation structure: a steering group, European Industrial Initiatives (EIIs), and the European Energy Research Alliance (EERA) and is supported by an information system (SETIS). Geothermal has been included in the SET Plan through the EERA Joint Programme on geothermal energy (EERA-JPGE) and the geothermal panel of the European Technology and Innovation Platform (ETIP) on Renewable Heating and Cooling whose aim is to develop strategic research priorities and implementation roadmaps for the sector. The launch of the Energy Union strategy, with its objectives to "make the EU the world number one in renewable energies", including by leading on the next generation of renewable technologies (European Commission, 2015, p. 16), has opened the door for even more focus on geothermal technology, including through a dedicated ETIP on deep geothermal energy launched in 2016.

8.4. State of Play and Evolution of National Incentives

All technologies pass through the same stages of the innovation cycle: from basic research through development, demonstration, deployment, and commercial market uptake. Instruments to internalize negative externalities of energy resources extraction, transportation, transformation, and consumption, for instance through a carbon tax or a carbon market, may not be sufficient alone to deliver the wide range of alternative and newer technologies at the necessary scale needed to decarbonize the economy. Where technologies are not yet competitive, direct (i.e., price or quantity instruments) and indirect support (e.g., favorable building codes, R&I funding) is justified, including for bringing costs down. Additionally, provided they are compatible with EU state aid rules, national mechanisms of support for RES are also specifically allowed by the RES Directive to achieve the national targets (Art. 3).

The promotion of one or more technologies in need of support is called "technology policy" (Linares *et al.*, 2013, 561–562]. Given its current limited market uptake and its significant potential, including in terms of cost reduction, an effective technology policy can change for the better the perspective of geothermal technologies in the energy markets. Indirect methods of support such as favorable building codes and R&I at EU level have already been presented in Sections 8.3.3 and 8.3.4. This section therefore focuses on the state of play and the evolution of direct financial incentives on the national level.

The level and type of support instruments for geothermal energy vary depending on the application, the market maturity as well the geological settings and the accessibility of the resource. For geothermal electricity, the main support is in terms of operating aid. Table 8.6 provides an overview of the operating aid available in 2016 in selected EU member states. The instrument of the feed-in tariff, a fixed and guaranteed price paid for each kWh produced, is considered the most attractive financial incentive for a project developer. As a matter of fact, the costs of capital for RES investments observed in countries with established tariff systems have proven to be significantly lower than in countries with other

Table 8.6   Operating aid for geothermal power in selected EU countries (2016).

Country	Type	Eligibility period (years)
Belgium (Flanders)	Quota system	10
Croatia	Feed-in premium	14
France	Feed-in Premium	15
Germany	Feed-in Premium	20
Hungary	Feed-in Tariff	N/A
Italy	Feed-in premium/Tenders	25
Portugal (Azores)	Feed-in tariff/Feed-in premium	12
Romania	Quota system	—
UK	Feed-in premium (contract for difference)	15

instruments that involve higher risks for future returns on investments. In the European Union, however, the new state aid rules for projects in the field of environmental protection and energy (EEAG) for the period 2014–2020 are phasing out feed-in tariffs in favor of more market-based incentives such as feed-in premium, i.e., a bonus on top of market price. This mechanism, depending on how it is designed, tends to increasingly expose renewable electricity producers to market signals. For geothermal, it has already been adopted in France, Germany, Italy, and the UK. Furthermore, the standard rule of the EEAG foresees that from 2017 this financial support should be allocated via a technology-neutral bidding process open to all technologies regardless of their maturity. Such a development may significantly increase uncertainty for less developed and more capital intensive technologies.

This is the reason why the following derogations apply if duly justified by member states:

- Feed-in tariff may be possible for demonstration projects.
- Member states may set up technology-specific bidding to ensure diversification, and take into account different levels of maturity.
- Support may be granted without bidding if the member states demonstrate that this would result in underbidding or in low project realization rates.

An alternative way to provide operating support to renewable electricity is through a quota system, which is a legal obligation on energy supply companies to purchase a specified amount of renewable energy. This instrument is used in Flanders (Belgium) and Romania and remains unchanged by the new state aid rules.

As far geothermal heat projects are concerned, public financial support is traditionally more fragmented and mainly allocated through grants covering part of the higher upfront investment cost compared to conventional technologies. In many member states, especially from Central and Eastern Europe, these funds largely stem from European Structural and Investment Funds. As pointed out by Connor *et al.*, (2013: p. 4) in their overview of support options for renewable heat technologies, "[g]rants can be easy to administrate and are attractive to governments wishing to stimulate initial interest [...] in particular technologies."

Some variations in terms of instruments to support geothermal heat projects is observed, for instance:

- grants combined with VAT incentives (e.g., France and Belgium) or with soft loans (e.g., France and Germany) with a guaranteed interest rate below market levels and with favorable repayment time;
- tax incentives only;
- operating aid similar to a feed-in tariff system like in the UK and, embedded in a multisectorial tendering scheme, in the Netherlands. Operating aid in the heat sector, however, is more complex and less popular compared to the power sector.

While of extreme importance, traditional financial incentives may not always be sufficient, especially for deep geothermal projects. How underlined in GEOELEC (2013, p. 50), "[m]ost of the investment falls into the high-risk phase of the geothermal project (Figure 8.3). While the project is being developed, the required budget changes successively. And with increasing effort in exploration, more and more knowledge about the resource is acquired and the risk of failure decreases accordingly." The bottleneck of many geothermal projects

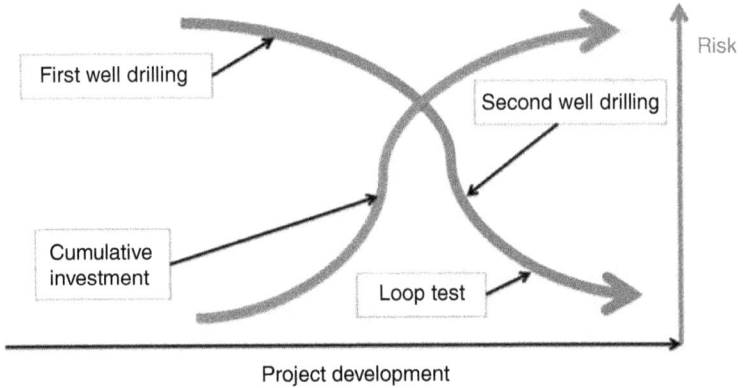

Figure 8.3   Risk and cumulative investment during the project progress. *Source*: GEODH (2014, p. 45).

is that in most cases "debt financing by banks is possible only following the completion of the long-term flow tests" (GEOELEC, 2015, p. 47). Furthermore, due to the limited practical geological knowledge in some regions, private insurers also consider the operation to be too risky. Under those conditions, a feed-in premium or a soft loan does not alone guarantee the successful financing of a project.

This barrier is a common issue and has been successfully overcome in France and the Netherlands where governments have taken action to set a public–private risk mitigation facility. As markets develop and costs decrease, in the medium-term, the private sector should be able to manage project risks, thereby enabling a sustainable long-term development.

8.5. Conclusion

This chapter has presented the key legal issues affecting the geothermal business (classification, resource ownership, licensing, sustainability, etc.) and has shown how EU policies and regulation impact the national and local frameworks. The EU environmental *acquis* ensures that a geothermal project is compatible with the environment, while EU energy and climate objectives for 2020 and beyond, and the accompanying financial and legal instruments put in place to facilitate their delivery can, in many ways, support the technology. In this context, the RES Directive has been the most important piece of legislation for geothermal energy. It addresses specific barriers for its market uptake, including the lack of a coherent definition and the persistence of burdensome administrative procedures. Moreover, EU directives promoting energy efficiency in buildings and products can facilitate the integration of geothermal for heating and cooling applications and increase awareness.

However, as shown by the case of national support schemes for electricity, policy, and regulatory instruments may evolve rapidly in response to new challenges. With the gradual phase-out of feed-in tariffs and with more constraints on public budget, the future of the sector may largely depend on innovative and hybrid financial mechanisms such as risk mitigation schemes coupled with more market-based incentives.

In conclusion, the recommendation for companies and organizations active in the geothermal business is not to underestimate the impact of these developments on their environment, but to rather follow and constructively contribute to the policymaking process at European, national, and local level.

References

Abate, S., Botteghi, S., Caiozzi, F., Desiderio, G., Di Bella, G., Donato, A., Lombardo, G., Manzella, A., Santilano, A., and Sapienza, A. (2014), VIGOR: Geothermal applications for a sustainable development. Production of heat and electric power (in Italian). CNR–IGG, ISBN: 9788879580120.

Angelino, L., Dumas, P., Nádor, A., Kepinska, B., Torsello, L., Bonciani, Lorenzen, S.B., and Kujbus, A. (2016), Regulatory frameworks for geothermal district heating: A review of existing practices, Proceedings of the European Geothermal Congress 2016, Strasbourg, France, paper. P-LA-308

Bradbrook, A.J. and Rønne, A. (2014). New advances in geothermal energy law: a comparative analysis. In D.N. Zillman, A. McHarg, A. Bradbrook, and L. Barrera-Hernández (Eds.), *The Law of Energy Underground: Understanding New Developments in Subsurface Production, Transmission, and Storage* (pp. 309–331). Chapter 16. Oxford: Oxford University Press. 10.1093/acprof:oso/9780198703181.001.0001.

Buchan, D. (2009), *Energy and Climate Change: Europe at the Crossroads*. Oxford: Oxford University Press.

Connor, P., Burger, V., Beurskens, L., Ericsson, K., and Egger, C. (2013). Devising renewable heat policy: Overview of support options. *Energy Policy* 59, pp. 3–16.

Delbeke, J., Klaassen, G., and Vergote, S. (2015). Climate-related energy policies. In Delbeke J., and Vis, P. (Eds.), *EU Climate Policy Explained*. Chapter 3. London: Routledge, (pp. 61–91).

Erhorn, H. and Erhorn-Kluttig, H. (2016). Nearly zero energy buildings: Overview and outcomes. In Concerted Action EPBD, 2016 — Implementing the Energy Performance of Buildings Directive (EPBD). (pp. 57–74). Adene. 978-972-8646-32-5.

European Commission (2011), Communication from the Commission to the European Parliament and the Council: Energy Efficiency Plan -COM(2011)109.

European Commission (2014), Energy Economic Developments in Europe, European Economy 1/2014, Directorate-General for Economic and Financial Affairs.

European Commission (2015), Communication from the Commission to the European Parliament and the Council: A Framework Strategy for a Resilient Energy Union with a Forward-Looking Climate Change Policy – COM(2015) 080 final.

Fraser, S. (2013), Report presenting proposals for improving the regulatory framework for geothermal electricity — Appendix I: Overview if national rules of licensing for geothermal. Available at: http://www.geoelec.eu/wp-content/uploads/2011/09/D4.1-A.1-Overview-of-National-Rules-of-Licencing.pdf.

GEODH (2014), Developing geothermal district heating in Europe, final report. Available at: http://geodh.eu/.

GEOELEC (2013), Towards more geothermal electricity in Europe, final report. Available at: http://www.geoelec.eu/.

Goldstein, B., Hiriart, G., Bertani, R., Bromley, C., Gutiérrez-Negrín, L., Huenges, E., Muraoka, H., Ragnarsson, A., Tester, J., and Zui, V. (2011). *Geothermal Energy*. In IPCC Special Report on Renewable Energy Sources and Climate Change Mitigation [O. Edenhofer, R. Pichs-Madruga, Y. Sokona, K. Seyboth, P. Matschoss, S. Kadner, T. Zwickel, P. Eickemeier, G. Hansen, S. Schlömer, C. von Stechow (eds)], Cambridge: Cambridge University Press.

Goodman, R. (2010) GTR-H — Geothermal Legislation in Europe, Proceedings of the World Geothermal Congress 2010, Bali, Indonesia, paper 0315, 1-5.

Hähnlein, S., Bayer, P., Ferguson, G., and Blum, P. (2013). Sustainability and policy for the thermal use of shallow geothermal energy. *Energy Policy*, 59(C), 914–925.

Hodson, P., Jones, C., and van Steen, H. (eds.) (2010). Renewable Energy Law and Policy in the European Union. Claeys & Casteels, Leuven. 9077644148, 9789077644140.

Linares P., Batlle, C., and Perez-Arriaga, I. (2013), *Environmental Regulation*. In I. Perez-Arriaga, (ed.), *Regulation of the Power Sector*, London, 2013, pp. 539–579.

Orbach, B., What Is Regulation? Yale Journal on Regulation Online 1 (2012), Arizona Legal Studies Discussion Paper No. 12-27.

REGEOCITIES (2015), Developing Geothermal Heat Pumps in Smart Cities and Communities, final report. Available at: http://regeocities.eu/results/

Tester, J.W., Anderson, B.J., Batchelor, A.S., Blackwell, D.D., DiPippo, R., Drake, E.M., Garnish, J., Livesay, B., Moore, M.C., Nichols, K., Petty, S., Toksöks, M.N., and Veatch, R.W. Jr. (2006), The Future of Geothermal Energy: Impact of Enhanced Geothermal Systems on the United States in the 21st Century. Massachusetts Institute of Technology, Washington, DC, USA. Available at: geothermal.inel.gov/publications/future_of_geothermal_energy.pdf.

Thieffry, P. (2015), *Traité de droit Européen de l'environnement*, 3ème ed., Bruylant, Bruxelles.

Index

www.ingramcontent.com/pod-product-compliance
Lightning Source LLC
Chambersburg PA
CBHW050543190326
41458CB00007B/1901